Y0-CBU-045

SCIENCE,
TECHNOLOGY,
AND
POLICY DECISIONS

About the Book and Authors

This text, written by a philosopher of science and a political theorist, introduces students to the issues and controversies surrounding science and technology policy in the United States. As the impact of technological advancement is increasingly felt, the policy-making process for science and technology is undergoing a marked transition. The making of this policy is no longer solely the function of government agencies and institutions. New actors in the policy arena are raising questions about the future of technological advancement in the United States and elsewhere, and their voices are affecting—sometimes obstructing—the traditional policy process. This book surveys the entire domain of science and technology policy making with special emphasis on the growing role of citizen participation, the ethical issues raised by modern policy problems, and the general principles that guide current policy.

The authors discuss current philosophical views about the nature of science and technology as social and political entities and also consider the history of the relations between these fields and political authority. They combine an issues and case study approach with a narrative discussion of how ethical, participatory, and institutional factors have merged in the policy process. Among the topics addressed are nuclear power and siting policy, hazardous waste, communications technology, and biomedical technology. After reviewing the difficult problems facing the modern policy maker, the authors assess the methods and ethical assumptions of the current policy-making framework and consider alternatives that are more sensitive to the complexity of contemporary policy issues.

Intended as a core text for courses in "Science, Technology, and Public Policy," the book can also be used in interdisciplinary courses focusing on the relationship between science, technology, and society. The text is also appropriate for courses in the philosophy of science and technology and for courses in social and political philosophy.

Anne L. Hiskes is associate professor of philosophy at the University of Connecticut, where **Richard P. Hiskes** is associate professor of political science.

SCIENCE, TECHNOLOGY, AND POLICY DECISIONS

Anne L. Hiskes
and Richard P. Hiskes

No Longer Property of
Phillips Memorial Library

Phillips Memorial
Library
Providence College

Westview Press / Boulder and London

Q
125
H629
1986

All rights reserved. No part of this publication may be reproduced or transmitted in any form or by any means, electronic or mechanical, including photocopy, recording, or any information storage and retrieval system, without permission in writing from the publisher.

Copyright © 1986 by Westview Press, Inc.

Published in 1986 in the United States of America by Westview Press, Inc.; Frederick A. Praeger, Publisher; 5500 Central Avenue, Boulder, Colorado 80301

Library of Congress Cataloging-in-Publication Data
Hiskes, Anne L. Deckard, 1951–
  Science, technology, and policy decisions.
  Bibliography: p.
  Includes index.
  1. Science and state. 2. Technology and state.
3. Policy sciences. 4. Science—Social aspects.
5. Technology—Social aspects. I. Hiskes, Richard P.,
1951– . II. Title.
Q125.H629 1986   306′.45   86-9276
ISBN 0-86531-631-7 (alk. paper)
ISBN 0-86531-632-5 (pbk.)

Printed and bound in the United States of America

∞ The paper used in this publication meets the requirements of the American National Standard for Permanence of Paper for Printed Library Materials Z39.48-1984.

10   9   8   7   6   5   4   3   2   1

CONTENTS

INTRODUCTION

Since at least the end of World War II, the United States has been a leader in the fields of science and technology. We are convinced that the same cannot be said of the United States in the area of science and technology policy, even though for forty years the U.S. government has spent billions of dollars in the pursuit of scientific and technological development. One of the purposes of this text is to examine and try to understand the reasons behind the failure of U.S. policy makers to provide a coherent policy for these two crucial areas.

Many books have purported to describe the policy process for science and technology, but in our view none of them has succeeded in identifying the diversity of actors and issues that contribute to this process. Policy making in these fields does not take place exclusively in bureaucratic agencies like the National Science Foundation or the Department of Energy; nor is it restricted to the Congress, the president, the courts, or agencies at the state level. Rather, the ranks of policy makers for science and technology have swelled to include U.S. corporate personnel, lobbyists, public interest groups, and private citizens with no affiliation.

The variety of issues perceived as relevant to science and technology policy is also expanding. As science and technology grow in scope and power, their possible effects on society and on individual lives become greater and more varied. The economy, national security, the environment, birth, and death are all radically affected by modern science and technology. But effects of this type raise difficult ethical issues about individual liberty, human rights, social welfare, and the just distribution of costs and benefits. These are precisely the kinds of issues that motivate public participation and heated controversy, and so they necessarily enter into the policy-making arena.

Our approach in this text is to examine policy making for science and technology from the diversity of vantage points that is required by the growing variety of policy actors and issues. Our discussions throughout the book deal with policy problems on several levels. On one level, we simply describe actions of people and institutions involved in the making of specific policies. On another level, we critically examine the current policy-making framework, calling attention to the many, often neglected, sides of science and technology policy issues. There are political, economic, technical, and ethical sides to each policy problem, and the actors in the policy-making process see these problems from the perspectives of different interests.

The many-sided nature of policy problems motivates us to ask two questions: Does the current policy-making framework adequately coordinate participation by many groups with varied interests? Does it adequately integrate economic, scientific, and political concerns with ethical considerations? Answers to these questions require us to consider fundamental philosophical issues that any policy-making framework for science and technology must ultimately address. These are issues about the nature of science and technology, about the proper role of lay citizens in directing these fields, and about the ethical principles that should guide policy decisions. By looking at science and technology policy from all of these perspectives, we hope to provide an accurate picture of the breadth and depth of modern policy problems and thereby to contribute ultimately to their solution.

In Chapter 1 we begin our examination of science and technology policy by considering how science and technology interact with each other and with society. We approach these questions from the perspective of the philosophy of science and technology, which addresses normative issues about the nature of these enterprises. A philosophy of science and technology is concerned with clarifying appropriate goals and methods for these activities. It identifies ideals toward which science and technology might reasonably strive and optimal methods for progressing toward these ideals.

As the historical anecdote that begins the first chapter illustrates, scientists, policy makers, and members of the public all implicitly hold some philosophy of science and technology. They have expectations about what these fields can and should achieve. As we maintain in Chapter 1, it is important for our policy-making framework to be based on an adequate philosophy of science and technology since a conception of these enterprises and their relations to society guides the kind of policy made and sets limits on who can make policy. In particular, we suggest that past policy has been based on an outmoded empiricist conception of science and technology that keeps the issues in these fields in the hands of "the experts" rather than in those of the public.

Chapter 2 offers a historical look at policy making for science and technology in the United States. Because it was founded during the eighteenth-century Age of Reason, this country has traditionally manifested a basic faith and optimism in scientific progress. But particularly in the period following World War II, political events and policy decisions have alternately challenged and revived this faith. Using the case history of the National Science Foundation as an example, we trace the emergence of a relationship between science and political power in policy institutions of the federal government.

In Chapters 3 through 8 we turn to specific issues and policy areas to demonstrate the complexity of science and technology policy and of the process that generates it. Within these chapters we employ a number of approaches and perspectives, all of which are necessary to grasp the breadth of the policy-making enterprise. We look first at the institutions of U.S. politics and policy making, such as the presidency, Congress, courts, and executive agencies to understand their essential roles in formulating science and technology policy. However, because much of what passes for policy in these areas is the direct result of political pressures, demands, and activity that occur outside these institutions, we cannot ignore the perspective of participatory politics and citizen involvement. The dramatic increase in citizen interest in issues like nuclear power, energy, and biomedical technology is evident to anyone who reads a newspaper or watches a television news broadcast.

As evident in the case studies that begin these chapters, the demand for citizen input into science and technology decisions is an example of how participation alters the definition of scientific and technological issues. Citizens view the issues covered in Chapters 3 through 8, not primarily as technical questions to be resolved by scientific experts but as issues that touch upon the traditional U.S. values of personal liberty, justice, and equality and upon the nature of democracy. A basic U.S. premise upheld even in court is that on issues of ethics and values, there are no experts. Therefore, many citizens are demanding that science and technology policy making be a participatory enterprise, and the need arises to examine policy making from one more vantage point—that of individual and social ethics.

In Chapter 9 we discuss specific ways in which the current policy-making framework might be revised to accommodate the participatory and ethical dimensions of modern science and technology policy issues. One thesis that emerges from Chapters 2 to 8 is that the current policy-making framework has failed because it does not incorporate the diversity of concerned actors and relevant issues into the beginning of the policy process. We therefore begin Chapter 9 by describing an analytical framework known as risk-cost-benefit analysis, which is commonly used in making policy decisions. As we see, this framework is typically applied in conjunction with assumptions of dubious ethical status. But, as we also show, this same framework can be used along with many different sets of ethical principles. The important point is that the public somehow be involved in the selection of these principles.

Perhaps the main theme of this text is that those who make and those who study science and technology policy need to be aware of the different perspectives that make up the policy environment. The philosophical, institutional, participatory, economic, and ethical dimensions of science and technology decisions must all be recognized and integrated. As our postscript points out, nowhere is this lesson more evident than in the tragic explosion of the space shuttle *Challenger.*

In our view, all citizens in a democratic society are, in the final analysis, legitimately called policy makers. This involvement is especially necessary in formulating science and technology policy, since all citizens must deal with the impacts of changes in these areas. Good policy makers must be sufficiently informed about the complexity of the issues and of the process in which they participate; this book is an attempt to inform those policy makers, both present and future.

1

SCIENCE AND TECHNOLOGY: PUBLIC IMAGE AND PUBLIC POLICY

On Christmas 1932 at the meetings of the American Association for the Advancement of Science,[1] newspaper journalists and over two thousand physical scientists packed the room reserved for a symposium on the nature of cosmic rays. The atmosphere was one of tense anticipation. Two Nobel Prize–winning physicists, Robert A. Millikan and Arthur H. Compton, were about to debate the merits of their opposing theories after two years of well-publicized rivalry. Millikan claimed that cosmic rays consist of photons produced by the fusion of hydrogen atoms. Compton claimed that cosmic rays consist of charged particles; in 1931 he launched a worldwide investigation to gather evidence for his theory, personally traveling fifty thousand miles over five continents.

The press was not disappointed with the debate. Millikan passionately defended his theory even though he had just received word privately that his anticipated confirmatory evidence was nonexistent. In a pique of temper after the debate, Millikan refused to shake Compton's hand. Headlines from New York City to Pasadena, California, carried the news to the eager public: The controversy remains unresolved—even Nobel Prize winners interpret data differently—even Nobel Prize winners are unable to live up to an ideal of dispassionate objectivity.

The level of interest shown by the press and by the public in the highly theoretical Millikan-Compton debate may seem puzzling. What was so important about the nature of cosmic rays? Why should the public care? The answer seems to be that public interest in the debate was motivated by growing doubts about the social value of science—

doubts already serving as fuel for a public revolt against science and technology.

The cosmic ray debate took place only three years after the beginning of the Great Depression. Millions of Americans were out of work, and they and their families were suffering great economic hardship. To the public, the Millikan-Compton debate epitomized the failings of science and technology: Not only had science and technology failed to prevent the Great Depression (and had, perhaps, caused it through mechanization and overproduction), but science appeared impotent to supply reliable knowledge. The high expectations for science and technology evident at the beginning of the twentieth century had been supplanted by a deep pessimism about their social value. As Robert M. Hutchins proclaimed in 1933 during his tenure as president of the University of Chicago, the nation was in despair because "the keys which were to open the gates of heaven have let us into a larger but more oppressive prison house. We think those keys were science and the free intelligence of Man. They have failed us."[2]

The tarnished public image of science and technology ultimately changed the face of research in the 1930s. Politically liberal critics blamed the failures of science and technology on the financial dependence of research on private wealth and industry. They believed that government policy was serving private industrial interests rather than public interests and that industries were using science and technology merely as an instrument to increase their own profits.[3] American Telephone and Telegraph (AT&T) was regarded as a major culprit, and in 1935 the Federal Communications Commission began an investigation of the company's policies toward research and development. Even philanthropic foundations, which played a major role in funding academic research, were not immune to criticism. The foundations, critics claimed, exerted a conservative influence on U.S. life by supporting only "noncontroversial" subjects like atomic physics and physical chemistry that could be "treated with scientific assurance." "Controversial" projects that might challenge existing values and policies, such as those involving the economic, environmental, or psychological effects of science and technology, were ignored. By ruling out projects with ethical implications, the critics asserted, foundations had "improperly forced the techniques of the natural sciences into the social sciences and the humanities" and had "overstimulated" certain fields such as physics.[4]

Many people influential in disbursing research money took the humanistic critiques of science to heart. By 1934 the officers and trustees of the Rockefeller Foundation had rewritten their previous funding policy: Now the only specific fields that qualified for support were those promising significant intellectual advances or those promising contributions to the "welfare of mankind," such as biology and psychology. In addition, the Rockefeller Foundation eliminated its previous $100,000 a year pledge to the National Research Council. As Daniel Kevles remarked in his history of this period, "not even Millikan could pry funds for cosmic-ray research out of the Rockefeller philanthropies, while at the same time Thomas Hunt Morgan, the Cal Tech Nobel Prize biologist, had no trouble winning ample help for research in genetics."[5]

The Rockefeller Foundation was not the only institution to change its funding policies: Other foundations followed suit, as did industry and the federal and state governments. By 1933, the National Bureau of Standards had laid off half its technical staff, and General Electric and AT&T had fired almost half their laboratory personnel. Nontenured university professors lost their jobs, and support for graduate students in the physical sciences was difficult to find. Research money for physics, chemistry, and engineering was virtually nonexistent. For a young person planning a scientific career, the physical sciences offered little hope for employment. Science was in a state of transition.

THE 1930S REEVALUATION OF SCIENCE

Widespread disillusionment with the direction of science and technology ultimately forced leaders in science, industry, and government to ask some hard questions about policy.[6] Issues raised by these leaders in the 1930s concerned the appropriate relations between science, society, and government. Should government take a more active role in funding research? If so, how should this role be defined? How much control was justified and by whom? Answers to these questions are as crucial today as they were in the 1930s. At issue is the formulation of a government policy that would be good for both society and the growth of science and technology. As the case of the 1930s shows, science policy can affect both the future course of science and technology and the future of humanity. The Rockefeller Foundation's shift in that decade from conducting research in physics to pursuing research in genetics turned out to be a crucial factor in the discovery of the genetic code,[7] and we owe our technologies of genetic screening and recombinant DNA to this discovery.

An additional lesson emerges from the case of the 1930s: Policy for science and technology is not spun out of thin air; it rests on prior assumptions about the nature of these enterprises. It is based on an image of how these activities best function and grow and of how science and technology interact. Prior to the 1930s, the frequently made assertions seemed to be that science and technology flourish best under a laissez-faire policy and that pure scientific research automatically translates into technological benefits. The Great Depression generated a great deal of skepticism about these assumptions. Furthermore, the case of the 1930s illustrates that assumptions about social values and their relation to science and technology necessarily guide policy. Prior to the Great Depression, attainment of the technological benefits of electricity, telephones, and an increase in consumer goods resulting from mechanization seemed like worthy goals of science and technology. But the relative value of such benefits later paled as the more pressing problems of employment and agriculture and the threat of war became priorities. Policy makers revised existing policy in part as a response to changed social values.

As a general conclusion, we may say that policy reevaluation is appropriate whenever public debate indicates that a discrepancy a exists between what the public believes science and technology should be achieving and what science and technology actually seem to achieve. This discrepancy can stem from the public's unrealistic expectations for science and technology, but it might also result from policy that fails to develop the full potential of these enterprises. Both factors seemed to play a role in the critiques of science and technology in the 1930s. During that time the public apparently believed that science should provide absolute and definitive answers to all important questions and that scientific research would automatically improve the quality of life. These expectations were naive in 1930, and they would be naive today. But, as we will discuss in the next sections, the problem may have been rooted equally in a policy-making framework that was based on an inaccurate conception of the nature of science and technology.

POLICY AND A PHILOSOPHY OF SCIENCE AND TECHNOLOGY

An optimal policy-making framework for science and technology would seem to be one that produces the best possible fit between social values and the internal dynamics of science and technology. An optimal framework, it would seem, must therefore be based on an accurate conception of the nature of science and technology. Policy should be based on realistic conceptions of the possible achievements of science and technology and of their relations to each other and to their social context. Positions on these issues are components of a philosophy of science and technology.

In the following sections, we will sketch two opposing philosophies of science and trace their implications for technology and policy making. The first position is logical empiricism; the second position is sometimes called "the new pragmatism" or simply "the new philosophy of science." We use the term *logical empiricism* to denote a philosophical framework that covers a class of philosophical positions that, although they differ on some points, agree on fundamental assumptions about science.[8] The picture of science as painted by the logical empiricists is probably familiar. From about 1900 to 1965 some form of logical empiricism was accepted by most scientists, philosophers, and historians as the framework within which scientific endeavor should, and in fact usually did, take place. Textbooks and popularizations of science during this period transmitted the logical empiricist view of science to the public. We assume, therefore, that many policy makers also saw science from a logical empiricist perspective. The "new philosophy of science" emerged during the 1960s within the fields of the history, philosophy, and sociology of science, and that approach is still controversial.

Although the philosophies of logical empiricism and the new pragmatism differ on crucial points, they share noncontroversial assumptions held by students of twentieth-century science. All accounts of science identify the acquisition of knowledge as the ultimate aim of scientific research and hold that a claim should be accepted as knowledge only when there are good reasons to believe that it is probably true or approximately true.[9] Furthermore, accounts of modern science agree that the primary test of truth within science is consistency with the data of experiment and observation. Scientists, however, are not only concerned with collecting data that describe particular, observed events. They also aim at discovering laws, like Newton's laws of inertia and of universal gravitation and Boyle's law of gases, that describe general patterns in nature. Scientists similarly construct theories, like Einstein's theory of relativity, Newton's theory of mechanics, and Darwin's theory of evolution. In this context a theory is a unified system of abstract concepts and general principles that serves to explain and organize a wide range of observable phenomena. Newton's theory of mechanics, for example, served to explain the orbits of the planets, the tides of the oceans, and the fall of an apple through the related concepts of gravity, inertia, and mass.

Although logical empiricism and the new pragmatism agree on all these points, their proponents employ different concepts of scientific truth and have different views on the nature of objectivity within science. Because of this disagreement, the two positions diverge on their characterization of the differences between science and technology and also on their accounts of the relations between science, technology, and society. In our opinion these differences have important implications for the connection between science and social and political values and for the legitimacy of public participation in setting research goals. The emergence of the new pragmatic conception of science and technology, we believe, calls for a reevaluation of policy goals.

LOGICAL EMPIRICISM: THE PHILOSOPHICAL IDEAL

The philosophy of logical empiricism has its roots in an old philosophical tradition. Since the time of the early Greeks, philosophers have tried to describe a foolproof method for achieving objective knowledge. Objective knowledge is supposed to consist of statements that are true in the sense that they accurately mirror elements of nature. The term *objective* is used when the content of knowledge is ultimately determined by the object under study and not by characteristics of the subject or observer. Objective knowledge is therefore uninfluenced by human interests or values; it is not relative to historical or social context. For the ancient Greeks, the most highly regarded form of knowledge was *theoria,* which meant "knowledge of the most fundamental, general principles of nature." These principles described the "hidden" causes behind observable phenomena. From the perspective of the ancient Greeks, only *theoria* could qualify as objective since only this form of knowledge could represent eternal, unchanging truths about the world. In contrast to *theoria* were the practical skills of the craftspeople and artisans. If these skills were ever elevated to the category of knowledge, they were considered to be inferior or vulgar forms since their goals were meeting practical, physical needs rather than attaining the ideal of pure objective knowledge.

In the eyes of twentieth-century logical empiricists, the ancient Greek distinction between *theoria* and practical skill is a valid one and characterizes important differences between science and technology. In their view, science aims at understanding nature—knowing what nature is really like—through the discovery of general laws and principles. In other words, a logical empiricist sees science as aiming at the discovery of objective, general truths. In contrast to science, technology aims at manipulating and controlling nature for some practical purpose through the creation of a physical tool or device.[10]

Logical Empiricist View of Science

A logical empiricist conception of science is characterized by three basic assumptions: (1) Data obtained through careful experiment and observation are objective; (2) there is one universally valid logic for science; and (3) through rigorous application of this logic to data, science gradually makes progress toward the ancient Greek ideal of *theoria.* The best way to clarify these assumptions is to see how they operate within the context of a typical logical empiricist account of scientific research.[11]

According to a logical empiricist, scientific research begins with a specific question generated by previously accepted scientific knowledge. A researcher may wonder about the causes of some observed event, about the laws that relate some given observed events, or about the more general, abstract concepts and principles that explain known laws. In each case accepted data motivate and mold the scientific question. In the second stage of research, a scientist formulates a conjecture (hypothesis) as a possible answer to the initial question. At this second stage, data again play an important role, since only hypotheses that are logically consistent with accepted data should be considered. At the third stage of research, new data are obtained through experiments suggested by the logical implications of the hypothesis. If the hypothesis contradicts the new data, then it must be revised or discarded in favor of a better hypothesis (that is, a hypothesis in closer agreement with the data). If, on the other hand, a hypothesis continues to agree with the data obtained through many different kinds of experiments done by many people, then eventually the scientific community accepts the hypothesis as accurately representing facts about the world. Once scientists accept a law or theory, they are not forced to support it forever; any law or theory may be revised or reinterpreted in light of future data.

According to a logical empiricist, although the laws and theories approved by the scientific community may change over time, the data should be accepted forever because data (that is, descriptions of what is observed) should be totally objective. Any rational observer of an event, it is assumed, must agree with the data describing the event. Data represent hard facts; they are given to the human observer by the world. Logical empiricists also assume that the logic employed by scientists has universal and eternal validity. In other words, the conclusions drawn by scientists about the adequacy of a claim or about the superiority of one hypothesis over another qualify as the most reasonable conclusions in light of the data. Any rational person in possession of the same data should arrive at the same conclusions.

Although no scientist would ever claim to understand absolutely the fundamental principles of nature, logical empiricists believe that scientific research makes progress toward this goal. Through a continuous process of proposing and evaluating hypotheses by means of logic and data, scientists obtain laws and theories that ever more accurately represent nature. To a logical empiricist, Einstein's theory of relativity may not be the final word on space, time, and gravitation, but it does provide a more accurate depiction of the world than did the earlier Newtonian theory of mechanics.

From the perspective of logical empiricists, the result of applying logic to objective data is the attainment of value-neutral scientific knowledge. By *value-neutral* knowledge, the logical empiricists mean that an accepted scientific idea simply depicts facts and that the idea does not embody or favor any particular human goal or value. Logical empiricists believe that the scientific method protects science from the prejudices and biases of individual scientists. More important, they believe that the scientific method insulates science from the changing fashions of political ideology and from the transitory pressures of economic need. Politics, economics, and social values are all potentially dangerous to the progress of science. Scientists who wish to find a result that would serve some special interest may collect only a certain kind of data, they may see connections among data that are not really there, they may reach premature conclusions about the acceptability of a hypothesis, and they may simply stop asking questions once their special interest is served. For science to fulfill its potential as a source for understanding the world, it must transcend its social environment, generate its own questions, and answer them by its own logic.

Logical Empiricist Approach to Technology

Logical empiricists distinguish between science and technology on the basis of the aims of those fields. Science aims at increasing our understanding of nature by discovering general laws, whereas technology aims at meeting some immediate practical goal by manipulating nature. Logical empiricists are not the only people who identify these aims for science and technology; most people define the differences between pure science or fundamental research and the other activities of applied science, engineering, and product development in terms of the same aims.[12] For our purposes, a significant implication of the logical empiricist position is that science should be objective, value-neutral, determined only by logic and data, whereas technology is inherently directed and molded by particular values and social forces. The logical empiricist view of the "value-ladenness" of technology agrees with most scholarly studies of technological change.[13]

Values and social context influence technology at the beginning of the research process when a problem is defined. A technological problem exists only in relation to human goals and social forces. A typical technological problem, for example, might be devising a machine that would produce a consumer product more quickly and within more precise specifications than currently possible. Presumably some individuals have adopted the goal of increasing monetary profits, and presumably increased production and precision are reasonable means for meeting this goal because of market forces. Second, human goals define what counts as a correct solution to a technological problem. If the machine "works" in the sense that it increases production and precision by an acceptable amount, then the machine is in fact a solution to the original problem. In science the situation is allegedly very different. A scientific claim is not acceptable simply because it might be useful.

General Acceptance

The intended distinction between science and technology, made by the logical empiricists, is embodied in the United States' language and legal system. Scientific advances are called "discoveries," whereas technological advances are called "inventions"; scientific discoveries are owned by no one, whereas technological inventions are owned by the patent holder. Philosophers of science see here an underlying presupposition that technology is a human creation designed to meet some human need, whereas scientific knowledge exists "out there" independent of human concerns.

A logical empiricist philosophy of science also seems to reflect accurately the self-image of many practicing scientists, particularly of physical scientists. The 1972 report by the National Academy of Sciences on the state of the physics community and its relation to society, for example, was written by prominent physicists who characterized their discipline in terms of logical empiricist assumptions: "Science is knowing. . . . If man is going to understand nature, he has to find out how it really is. There is only one way to find out: experiment and observe. . . . What is learned in physics stays learned. . . . Each change [is] a widening of vision, an accession of insight and understanding."[14]

Robert Merton, a prominent sociologist of knowledge, identified a set of norms that he claimed were publicly espoused by most scientists from 1600 to 1970. These norms, as they are described by Robert Ackermann, reflect the logical empiricist goal of a value-neutral science.[15] Scientists claimed to be impartial because they were "interested in pure knowledge, not in applications." They claimed that personal rewards were to be achieved through "community recognition" and that "financial gain through discovery is wrong." Empirical evidence was also identified as the "only arbiter of truth," and "secrecy about results was considered immoral." Ackermann's conclusions were that "these norms portray the picture of science that many scientists prefer to see as the public image of their profession" and that "this set of norms can be thought of as standing to science as its ideology."[16]

THE PHILOSOPHICAL IDEAL: POLICY IMPLICATIONS

Imagine a federal policy maker who adopts the images of science and technology sketched in the preceding section. When he or she is deciding upon the most appropriate type of policy concerning science and technology, one implication would be clear: If science and technology are very different types of enterprises, they probably would require different types of policies. In this section we will first explore policy options compatible with the philosophical ideal of scientific knowledge, and second, we will examine policy options for technology. As we will see, the formulation of a particular policy depends on the policy maker's perception of the public good and also on the policy maker's assumptions about the interaction between science and technology.

Policy Options for Science

According to the logical empiricists' image of science as concerned with the production of objective knowledge, science must be protected from external social forces in order to function successfully. Scientists must be free to pursue interesting scientific questions as they arise, choosing them because of their potential contributions to the body of scientific knowledge. Furthermore, scientists must select answers to these questions on the basis of data and logic and not because the answers serve some political ideology or some economic or social value.

Government Funding. One aspect of science policy, therefore, could be to provide protection for science from external social forces. At most, government may supply funds for pure research. The scientific community should be in charge of disbursing these funds on the basis of scientific merit, presumably through a panel of experts chosen from all scientific disciplines. The government should not target specific research topics or approaches for preferential treatment, and lay participation in choosing research topics is entirely inappropriate.

Some scientists greet any proposed link between science and government with suspicion. Robert Millikan of the famed Millikan-

Compton cosmic ray debate feared any government funding of pure research. Throughout the 1930s, Millikan and others "remained steadfast in their faith in the social wisdom of relying upon the private sector."[17] Millikan's position was not without foundation. The case of genetic research in the Soviet Union during the 1930s and 1940s illustrates the potential dangers of governmental interference in the scientific process.[18] T. D. Lysenko, who was the governmental head of Soviet genetic and agricultural research, allowed only research based on Jean Baptiste Lamarck's theory of evolution. (Simply stated, Lamarck's theory of the inheritance of acquired characteristics implied that an organism can be genetically altered during its lifetime by changing its environment; this is in contrast to Darwin's theory of the survival of the fittest, which implied that a species can be genetically altered only over several generations of selective breeding.) Lysenko claimed that only Lamarck's theory was compatible with the Soviet ideology of improving the human condition through environmental change. Because Lamarck's theory was not adequately supported by data, Lysenko's policy had disastrous effects on Soviet agriculture and set back Soviet biological research by decades. The Lysenko case is admittedly an extreme example of governmental interference in science, but it illustrates what can happen once government begins directing science. Even the limited role of government as a major supplier of research funds could stifle research with bureaucratic red tape.

Although Millikan and like-minded scientists have reasons for fearing governmental intervention in science, one can argue that even they must admit that governmental funding is the lesser of two evils (the other being private funding). Modern scientific research is expensive. In nuclear physics, for example, the approximate cost of operations and equipment to support the research of one Ph.D. scientist for one year is $80,000.[19] The money must come from either private or public sources. Private industry and private foundations are free to adopt any funding policy they wish, subject entirely to the personal preferences and the values of those in control. The freedom and objectivity of twentieth-century science seem much better served under an ideal system of governmental funding in which appropriate safeguards for impartiality and disinterestedness may be imposed with legal force.

Payoff of Scientific Research. We have seen that governmental funding of pure research is compatible with the philosophical ideal of scientific knowledge as long as government acts simply as a source of money. But now the policy maker faces an important issue: Public money should be spent on projects of clear social value and it should be budgeted in proportion to the social importance of the project. What justification can be given for funding pure scientific research, particularly if the money spent on science far exceeds that spent on the arts and humanities, or on social programs? The attainment of scientific knowledge is undoubtedly a worthwhile endeavor, which contributes to our general cultural enrichment. But is there any additional social payoff from pure research? If so, the social payoff must come from science-based technology.

Since 1945 policy makers have typically justified government support of pure research in terms of what historian George Wise called "the assembly line model" of science and technology.[20] According to many policy makers, the assembly line model accurately represents the relation between twentieth-century science and technology. The model, as described by Wise, locates the origin of all modern technology in prior scientific knowledge, and it explains the development of technology in terms of an "assembly line" of stages beginning with the initial conception of a scientific idea and ending with some consumer product. The model identifies research for the sake of knowledge with science, and it identifies research for the sake of consumer products with technology. The assembly line model, therefore, is compatible with a logical empiricist philosophy of science and technology.

Wise depicts the assembly line model of technology as follows, where the arrows should be interpreted as representing a one-way process of development.

Scientific Idea —>Scientific Research —>Knowledge —>Applied Research   —>Invention   —>Development   —>Engineering —>Marketing[21]

If they assume the accuracy of the assembly line model and the social value of technology, policy makers can justify spending large amounts of money on pure research in the interests of promoting technology.

Policy Options for Technology

Unlike the case of science, social forces and practical goals always determine the current state and direction of technological research. Because technology inherently serves the purposes of some individual or group of individuals, government has more flexibility in developing a policy for that field than for science. The policy maker's approach is to determine the best way to direct and stimulate technological innovation in the public interest. The possible policy options range from a free-market, laissez-faire approach to that of complete centralized planning. The only restriction for technology policy imposed by the philosophical ideal of scientific knowledge versus applied technology is the requirement that governmental projects seek to apply already existing knowledge. Violation of this requirement constitutes governmental interference in the scientific process.

Laissez-faire Policy. On one end of the technology policy spectrum, a policy maker may consider a laissez-faire policy, defined as a policy under which the market forces of supply and demand alone direct technological growth. According to this policy, private industry would be the major source of funding for applied research, and the profit motive would be the primary reason for funding this research. Members of the public would influence the direction of technology by choosing to buy one product rather than another. (Imagine how long technological innovations in the computer industry would continue to be made if very few people bought any type of computer.) The government, as a major buyer of devices with potential military applications, also would influence the direction of technology.

Under a laissez-faire philosophy, whose advocates include political conservatives like the well-known free-market economist Milton Friedman, the role of government in relation to technology should be limited as much as possible to that of a buyer. According to laissez-faire principles, government may legitimately influence private business and industry only in the interests of protecting individual rights granted by the U.S. constitution, and these rights must clearly take precedence over the right of other individuals to conduct their business as they see fit. As an illustration, government may regulate the pollution levels of industrial smoke stacks only if the right to clean air takes precedence over the right of industrialists to maximize their profits.

Critics of a laissez-faire policy for technology charge that it does not always work in the best interests of the public. If profit is the primary motive behind technological innovations, innovations with limited market value would simply never be developed. Consider the plight of people suffering from a rare disease. The initial costs of developing a cure for the disease may be great, and if the disease is extremely rare, the market for the cure would be very small. Cost-benefit ratios would make it irrational for a business to underwrite the project. In general, technology would probably never be developed to solve a problem shared by a small minority with little money.

Under a laissez-faire policy, one would also expect a large number of conservative technological innovations of minor social significance, as opposed to a few radically new innovations designed to meet important human needs. As many studies show, increased profits obtained by a firm from extensive research and development rarely outweigh the costs of that research and development, particularly when competition among firms is great.[22] Given the profit motive, an industry would aim at minor revisions in a product with a known sales record rather than sink large amounts of money into applied research for a radically new type of product. In the latter case, a business would run the risk of producing a product that no one wanted to buy. A competitor could also steal the market before a business gets its invention into production. Consequently, industries would typically produce new, improved versions of old consumer products, or they would aim at producing old products in larger quantities and with more cost-effectiveness. As a further consequence, one would find industries' duplicating each other's research efforts for the sake of producing an item virtually indistinguishable from that of their competitors. This type of innovation, the critics claim, underutilizes human ingenuity and the full social potential of technology.

National Policy. Critics of a laissez-faire technology policy were particularly active in the years after the Great Depression. Competition and overproduction, they claimed, had caused the economic collapse of the nation. Even corporate leaders began "generating plans for national planning."[23] National goals, they believed, required businesses to coordinate their efforts. Gerald Swope, president of General Electric and a respected governmental policy consultant, recommended that "all industries involved in interstate commerce with over fifty employees should organize into trade associations with the power to regulate prices and consumption. In exchange . . . , labor would be guaranteed employment or unemployment insurance." Karl Compton, brother of Arthur Compton and chairman of the federal Science Advisory Board under President Roosevelt, extended Swope's ideas to include a long-term national plan for all applied research.

The vision of Swope, Compton, and others—a large-scale, long-term national plan for technology—represents a policy option opposed to the laissez-faire policy in the extreme. It opens the door for increased public participation in the policy process: The public, perhaps through its elected representatives in Congress, could formulate a priority ranking of desired products, and government would then implement a plan in line with these priorities.

Within the current U.S. framework, a natural mechanism for achieving national goals is the distribution of tax dollars or tax incentives with strict stipulations attached. For example, if producing drugs for rare diseases is identified as a national goal, then $10 million may be set aside for funding such projects. If the production of more or better computers is assigned a high positive ranking by the public, while the production of improved television sets is assigned a low ranking, then computer industries may be subsidized and television industries penalized. The technological projects considered here focus on a particular kind of device with a definite, foreseeable use. Federal funding of general targeted areas of research, such as space exploration or the causes of cancer, is undesirable because discoveries in these areas would require pure research, thus violating the ideal of the separation of science and government.

Directing the course of technology according to a national plan has several advantages. Goals requiring a collective effort and/or vast amounts of money would have a greater chance of being reached than they would under a laissez-faire policy. Concentrating efforts on select areas would increase the probability of ground-breaking developments in a shorter time. Furthermore, the idea of national planning based on common interests carries with it all the possible advantages of conscious, future-directed, rational deliberation. Policy makers can weigh the long-term and short-term consequences of various courses of action, they can identify and deal with needs before they become acute, and they may consider the whole picture of the mutually interactive parts of technology and society. In particular, personnel needs over a long period can be coordinated with the educational system so that the necessary people can be trained before they are required. People would be able to choose careers with reduced risks of unemployment. The ideal outcome of this planning approach when applied nationally would be that the public, through its representatives, would exercise complete control over the development of technology and would use it to the best and surest advantage of society.

Although the approach of directing technology according to a comprehensive national plan is attractive, it also faces inherent difficulties. One notorious difficulty concerns the identification of the public good. Must the public good benefit everyone or only a majority? Can the public good be determined by what people say they want, or are people generally poor judges of their own best interests? A second difficulty concerns predicting the long-term effects of using a technological product. Harmful side effects of drugs like aspirin or of some fertilizers become apparent only after many years of use. If people use these products in large quantities, then the possibility of harmful side effects occurring is great. Furthermore, concentrating on a relatively small number of products to the exclusion of others limits the adaptability of a society to changing needs and circumstances. Although large-scale technological planning that concentrates on select projects offers potentially great rewards, it also offers potentially great disasters. Perhaps the wisest course for a policy maker is to hedge his or her bets by adopting some kind of middle course between a laissez-faire policy and a complete national plan.

SCIENCE AND TECHNOLOGY: THE NEW SYNTHESIS

Many members of the scientific community concur with the logical empiricist image of science as providing objective knowledge of natural laws. This image enables the scientist to defend his or her turf from intrusions by government or the public. The assembly line model, which establishes the connection between science and technology, endows science with great social value and provides scientists with ammunition in arguing for public funding. Thus an argument based on the logical empiricist ideal of objective knowledge—combined with the assembly line model of science and technology—can be used to legitimize placing large amounts of public money in the hands of the scientific community.

Accuracy-Inaccuracy of Assembly Line Model

Careful examination of the historical record, however, casts strong doubts on the accuracy of the assembly line model for science and technology. A growing number of historical studies indicate that most technological innovations result primarily from the efforts of applied inventors and engineers.[24] New discoveries made by "pure" scientists of general laws of nature have contributed only minimally to technological inventions. The knowledge needed to create a new product or new process for making an old product is typically generated by the inventor. Even the field of microelectronics—which is regarded as the embodiment of high technology—owes its existence to the technological, not scientific, community. The point made by these studies is that technological innovation does not always depend on prior scientific knowledge; inventors frequently create sophisticated new tools without a deep prior understanding of why the tools work.

Historians are not the only skeptics about the assembly line model by which science feeds technology. Government policy makers, who began questioning the model in the 1960s, wondered whether the vast amounts of money spent on pure research were really paying off. Motivated by such doubts, the Department of Defense sponsored Project Hindsight in the 1960s to determine "the payoff to Defense of its own investments in science and technology."[25] After identifying crucial events that had led to the development of twenty military systems, the study concluded that only 0.3 percent of the these events resulted from undirected, pure scientific research. The study also concluded that such research is of great value but only on a "50 year or more time scale."[26]

The debate about the social value of pure science remains unresolved. However, if policy makers reject the assembly line model of science feeding technology, then the scientific community is confronted by a dilemma: Either it must face increased difficulty in competing for public money, or it must forge a new synthesis with technology.

HISTORICAL APPROACH TO SCIENCE AND TECHNOLOGY

A new image for science that paves the way for a close relationship with technology emerges from recent work by historically oriented philosophers of science and technology. Rather than beginning with an ideal to which scientific practice should conform, historically oriented philosophers begin by examining actual science as it has been practiced over the centuries.[27] On the basis of historical evidence, these philosophers conclude that science has changed over time: It has utilized different logics, pursued different values, and rejected the data collected by previous generations. These changes in science have been accompanied by changing relations between science and technology. These conclusions are significant for our purposes because they suggest that the logical empiricist conception of science may be in error: There may be no unique, universally valid logic for science; there may be no such thing as a value-neutral science; and even data may not be totally objective facts. If the nature of science and its relation to technology change over time and are always relative to social context, then perhaps the ideal of scientific objectivity should be discarded as an obstacle to a new synthesis between science and technology.

The changing relations between Western science and technology described by historians may be characterized very roughly in terms of four historical periods: (1) ancient Greece through the fifteenth century, (2) the Scientific Revolution from about 1540 through 1750, and (3) 1750 to 1940. The beginning of a fourth period in 1940, which would extend through the present, is somewhat speculative, and its classification as a distinct period depends on the actions of the scientific community and on future science and technology policy.

Ancient Period. During the first period the practices of science and technology were pursued independently of each other. Technology consisted of arts and crafts like metalworking, architecture, and medicine. Science was a branch of philosophy, and the formulation of scientific theories proceeded for the most part according to the a priori method with little place for experimental testing or the trial and error empirical method of technology.

During the Middle Ages all scholars, of whom Thomas Aquinas is a representative example, organized their world according to the ancient Greek philosophy of Aristotle. Aristotelian principles provided a dogmatic, conceptual framework for Catholic theology, science, ethics, and politics, and within the Aristotelian framework all these areas were integrated into one organic, natural order.

Scientific Revolution. Sometime during the sixteenth century, the scientific revolution associated with Copernicus (1473–1543), Galileo (1564–1642) and Isaac Newton (1642–1727) was born. During this revolution, scientists adopted a new method of inquiry—the experimental method advocated by Francis Bacon; they also took on a new system of principles for understanding nature—the mechanical philosophy advocated by Descartes and Galileo. By accepting the mechanical philosophy, scientists rejected the earlier Aristotelian unity of science, theology, and ethics. From the perspective of the mechanical philosophy, the principles governing the material, physical world are entirely different from those governing the spiritual or mental world. To understand the physical world, say advocates of the mechanical philosophy, scientists must interpret phenomena in terms of particles of matter exerting forces of push and pull. Aristotle's principles may apply to theology and ethics, but they do not apply to matter.

By accepting the experimental method as the special method of scientific inquiry, scientists separated the methods and results of science from the methods and results of theological and ethical inquiry. In science the only test of truth is agreement with data, as opposed to agreement with some authority like Aristotle or the Catholic Church. Many factors contributed to the Scientific Revolution: Some were integral to the field of science itself, but many were the results of the changing political, social, and religious environment. Authority and dogma were being challenged on all fronts as people came to see freedom of belief as an important value.

The relations between science and technology during the Scientific Revolution (from about 1540 to 1750) were very intimate. The two disciplines shared a common method—the empirical approach of analyzing correlations among data—and a common conception of the nature of their achievements. Most scientists during this period identified the collection and organization of data as the only legitimate aim of scientific inquiry. The creation of abstract, all-encompassing theories and principles that described the "hidden" causes behind the data and that served to explain the data was not the job of a scientist but of a philosopher. Scientists should not make inferences that go beyond the data but should be content with arriving at propositions like "heating a mixture of charcoal and ore produces a metal," which describe relations among observable events and properties.

During this period scientific and technological research produced essentially the same type of knowledge, and scientific truth was equated with utility. Knowledge that two types of events are correlated was usually the only type of knowledge necessary to produce a device useful in manipulating nature. Thus claims accepted as true by scientists had potential utility. Conversely, the ability to manipulate nature by causing one event that produces a second event is the ultimate experimental test of a proposed correlation. Since the only truths to which science could aspire were claims about correlations among observable events, claims useful in manipulating nature could be legitimately accepted as true. Francis Bacon's words "knowledge is power" characterized the operative philosophy of science and technology during this period. As a result of this philosophy, Bacon and others "took a deliberately strong position against the idea of a separation and opposition between technics and science, manual and intellectual work, and mechanical and liberal arts."[28]

The shared philosophy of science and technology brought about close working relations between scientists and technologically oriented researchers. Unlike the picture presented by the assembly line model in which technology follows science, during this period science followed technology. The useful devices of technology provided material for scientific analysis, and technological problems motivated scientific research. "The pumping of water, . . . the ballistic curves of projectiles, the optimal shape of ships' hulls, etc., were incorporated into the concern of science."[29] The social institutions of science and technology, which at the time consisted of special academies and societies, also reflected the close connection between the two disciplines. The Italian Academia del Cimento, the French Academy, and the British Royal Society incorporated the Baconian ideal of the unity of knowledge and utility in their charters, and the memberships of these institutions were made up equally of scientists, inventors, and engineers.

1750–1940 Period. The unity of scientific truth and utility did not last. In the early 1700s scientists began shifting their interests from ordinary, everyday objects to more abstract, hidden causes. Atoms, molecules, and gravitational and magnetic forces became the relevant objects of scientific inquiry. Once again scientists became oriented toward the formulation of theories only remotely connected to the data of everyday life. Practical issues no longer generated scientific questions, and practical utility was no longer a measure of scientific value. Scientists adopted the philosophical ideals of universal, explanatory principles as their guiding objectives. It is probably no accident that the philosophy of logical empiricism was articulated and widely accepted during this period of increasing theoretical activity.

During the 1750–1940 period the growing philosophical separation of science and technology became institutionalized. Two distinct communities emerged: the scientific community and the technological community. The communities worked and developed in isolation from each other, each having its own set of problems, its own criteria for evaluating solutions, and its own network of communication between colleagues. Furthermore, each community generated its own body of relevant knowledge; technology was not applied science. As a consequence, the educational processes of science and technology also were separated. Scientists worked primarily at universities, institutions in which research and education could be combined. Technical schools took over the training of engineers and other technical personnel who would later be employed in business and industry. The institutional separation of science and technology may have contributed to the growth of science: It allowed scientists to concentrate on increasingly more abstract and complex problems without the distracting concerns of applications and possible consequences.[30]

Although the scientific and technological communities were isolated from each other during the 1750–1940 period, as they perhaps are even today, occasional transfers of knowledge between them did occur. Such technologically motivated inventions as the telescope, microscope, and laser have become very useful instruments in scientific research.[31] Conversely, scientific discoveries such as Dmitry Mendeleyev's periodic table of the elements and Friedrich Kekule's theories of chemical compounds provided a foundation for the first science-based industry.[32] Transfers of knowledge like these are rare; more typically the scientific community invents its own research instruments, such as the air pump and the particle accelerator. [33] Likewise the technological community usually generates its own knowledge before it is applied to solve a practical problem.

A New Age of Science and Technology

Some historians and sociologists of science claim that the middle of the twentieth century marks the beginning of a new age for science and technology, an age during which these disciplines will once again be united under the Baconian ideal of the unity of knowledge and utility.[34] Unlike the first Baconian age (1540–1750) in which technology led science, this new age will be characterized by the "scientification of technology"; however, it could just as well be called the "technologization of science." During this period, scientists and technologically oriented people will work together in teams to find solutions for very practical problems. New lines of communication between the scientific and technological communities will open; members of the two communities will share problems and results. Perhaps any meaningful distinction between the two communities will eventually disappear.

Current trends in science and technology are compatible with the vision of a second Baconian age. Because scientific theories are becoming more complex and abstract, more sophisticated machinery is required for their testing; as scientists are uncovering new research areas new instruments and techniques are needed to explore them. Today scientists must not only invent, maintain, and improve their own instruments; they must also be skilled in operating them. At the same time that science is requiring more technological input, technology is moving more into the realm of science. Complex, abstract technical problems require abstract principles for their solution, and novel types of problems typically require a period of basic research before they can be solved. The recent growth patterns of science and technology therefore indicate that their practices may be converging.

Further signs of a unification between science and technology are evident within business and industry. An increasing number of industries employ teams composed of both scientists and engineers who share a common task of meeting some industrial need. The phenomena of big industrial research laboratories and of the research entrepreneur who directs them are fairly recent developments. A research entrepreneur is "an individual dedicated to creating new science or new technology, but realistic enough to recognize that he must strike bargains with people who have very different interests if he hopes to accomplish his goals."[35] As director of a large industrial laboratory, like that at General Electric Company or Bell Laboratories, the research entrepreneur is forced to compromise between the needs of the industry and the ideal of research for the sake of increased knowledge. Because the aims of science and technology seem so different, historian Melvin Kranzberg called these combinations of science and technology a "marriage of convenience," not a "love match."[36] Whether the scientific and technological communities will ever really become one community depends in part on future science policy and in part on the attitude of scientists toward research directed from outside the scientific community.

The image of science based on the logical empiricist ideal of objective truth, undistorted by external social factors, is a major obstacle to a complete union between science and technology. As long as their ultimate goals of abstract knowledge and practical utility are seen as inherently different, science and technology must regard each other only as a resource for occasional use. Contemporary philosophy of science, however, provides a new image for science that dissolves this obstacle. Many contemporary philosophers of science claim (1) that scientific knowledge on the levels of both explanatory principles and data is never really objective and (2) that, in the final analysis, values other than the abstract value of truth always direct and mold the content of accepted scientific claims.

The basis for these two philosophical claims may be argued as follows. The history of past and present science, contemporary philosophers of science claim, reveals that science always has been and always will be molded and directed by the values of human beings: There is no other alternative. During each historical period, the scientific community operates with a set of values and norms that define what problems are legitimate scientific problems and what counts as an acceptable solution. Logic and data alone are never sufficient to define the worthiness of a theoretical problem or to determine a uniquely correct solution. Personal and community values thus necessarily supplement logic and data.

Furthermore, the philosophers of science continue, the values of explanation, prediction, precision, reproducible data, generality, and mathematical elegance have all played a role in the evaluation of scientific theories, although differing levels of importance have been attached to these values during different historical periods. As the values change, so do the activities and results of scientists.

Most important, however, the character of scientific logic and data also changes with changing values. Scientists, like all people, necessarily perceive the world with an interpretive system of concepts and presuppositions. The interpretive system directs the scientist to note some features of events while ignoring others, and it provides the vocabulary by which data are described and eventually organized into patterns. The interpretive system also provides links between concepts and statements, and these links supply the logic by which a scientist moves from one statement to another. In some very literal sense, it is claimed, accepted dogmas and values always influence what scientists observe and how they reason.

According to this new conception of science, all accepted scientific knowledge contains a subjective component. Scientific truth is whatever claims best conform to the "scientific" values of a majority of scientists at a particular time. Scientific truth may then be viewed as equivalent to utility in a broad sense of the word: Scientific truth is what at the time is most useful in achieving the goals set for science.

Science is not molded and directed only by some value system adopted by the scientific community; external social values also necessarily influence science. The values and norms of the scientific community change over time, and at any given moment alternative systems of values are logically possible. So why does the scientific community adopt one set of values rather than another, and what factors bring about a change in these values? One plausible explanation is that the values of scientists, like the values of any other people, are influenced by the cultural values of their day. Furthermore, the scientific community, like any other professional group, probably adapts its behavior to social expectations, social needs, or a social level of tolerance. Consequently, one would expect changing norms and values of the scientific community to bear some correlation to changing religious, political, and economic values of the surrounding community. Furthermore, if transcendent, objective knowledge in science is just a delusion—because in the final analysis such knowledge can be reduced to some form of utility—then it is no longer imperative to isolate science from the influence of society at all costs.

In this section we have seen that many studies from 1960 to the present in the history and philosophy of science suggest a new image for science and technology. In the new image science and technology are depicted as dynamic social institutions whose form and inter-relations change over time. Each institution evolves in part according to its own internal logic and in part as a response to its wider social environment. As some historians predict, the institutions of science and technology may now be converging toward one common set of problems, methods, and internal values. If recent philosophical critiques of science are correct, then the unification of science and technology cannot threaten a scientific objectivity that never really existed.

THE NEW SYNTHESIS: POLICY IMPLICATIONS

Imagine a federal policy maker who consciously adopts an image of science and technology based on the historical and philosophical material presented in the previous section. How will the approach of this policy maker differ from that of a policy maker who accepts the image of science and technology patterned after the philosophical ideal of abstract scientific knowledge obtained in isolation from outside influences? Our hypothetical policy maker rejects the assembly line model in which technology supposedly feeds on undirected scientific research. He or she simply regards undirected, pure research as ineffective in yielding short-term technological benefits. Consequently, our policy maker cannot justify spending large sums of public money on undirected scientific research without another defense.

Considerations in Choosing a Policy

Even though our policy maker rejects the assembly line model of science and technology, he or she has other options. Government direction of scientific research is perfectly acceptable to this policy maker because the ideal of objective scientific knowledge has been dismissed. In its place we have knowledge formulated by the scientific community according to its own system of subjective values, which to some degree already reflect wider social values. Our policy maker therefore sees nothing wrong with government's acting as a social force upon the values and output of the scientific community, particularly in the name of the public good. Science need not be jeopardized as a consequence.

A policy maker assuming the new, historically based image of science and technology has more flexibility in developing a legitimate policy. He or she may allow government to target specific basic research areas or approaches for special funding, perhaps even eliminating public support for an entire field. Funding decisions can be based on the projected practical applications of the research, and they may affect the future of science: Because contemporary scientific research is expensive, out of necessity scientists and science students would gravitate toward fields with money.

By adopting the historical approach our policy maker may also consider ways of effectively combining science and technology. The current highly complex state of technology requires the guidance of reliable scientific knowledge for innovations and improvements to continue. Improving the design of nuclear reactors and safely disposing of nuclear waste are problems of the 1980s that can no longer be solved by the previously harmless technological method of trial and error tinkering: Action in such areas must be based on reliable principles governing the relevant processes. Many scientific problems of great social concern are also best addressed by a strategy that combines science and technology. Developing a cure for cancer, for example, requires the directed efforts of both basic and applied researchers.

Science has therefore become a valuable and necessary resource for technology, but as long as the "pure" scientific community operates in isolation from the "impure" technological community, this resource is not effectively used. Under the historically based image of science and technology, our policy maker is free to develop policy that would help merge the scientific and technological communities. He or she may legitimately explore new avenues for coordinating the efforts of the two communities, perhaps using the example of an industrial research laboratory directed by a research entrepreneur as a model for national laboratories.

A policy maker who adopts the historically based image of science and technology faces deeper and more difficult issues than a policy maker who regards science as isolated from social concerns and influences. Because the former policy maker believes that social forces always influence the direction and outcome of research, he or she bears the responsibility of selecting, and even of creating, the operative kinds of forces. But what values should he or she use in making this selection, and whose values should be given priority? These questions are particularly pressing in light of the potentially great scope and power of a combined science and technology. But what kinds of concerns should be given priority?

Furthermore, scientific knowledge combined with technical expertise is a powerful tool for creating new ways of manipulating and controlling nature with greater and longer lasting effects. The consequences of manipulating the energy of the atom or of altering the genetic heritage of the human species, for example, are of a new order of magnitude: they affect life and health over the long term, as opposed to consequences that merely affect one's level of material comfort. Because the social consequences of a science-based technology are potentially quite dramatic, a larger variety of social values are now potentially relevant to a decision about the course of science. Our policy maker can no longer regard science as simply the search for factual truth: He or she must consider how scientific knowledge interacts with political, ethical, and even religious values.

Possible Policy Approaches

The available policy options regarding science-technology range anywhere from a laissez-faire policy to a complete national plan. On the surface, a laissez-faire policy may seem an easy way for a policy maker to avoid difficult issues about the interrelationships among science, technology, and social values. By choosing a laissez-faire policy, a policy maker would select the free-market forces of supply and demand as the dominant social forces acting on both science and technology. Under this policy, private business and industry would be the primary sources of research funds, and research would be conducted on the basis of industrial interests ultimately geared toward the consumer. By choosing a laissez-faire policy, a policy maker would decide that the primary value relevant to science policy is the value of free enterprise.

The shortcomings of a laissez-faire policy, which we have already discussed in connection with a policy for technology, apply to science policy as well. An industry would be unlikely to support research that does not translate into increased profits for itself. Although science and technology could be effectively combined to meet socially significant national goals, a laissez-faire policy—because it would define scientific research in terms of industrial interests—is particularly wasteful of the social potential of science.

Perhaps the most important policy implication of a historically based image of science and technology concerns citizen participation in the policy process. Public concerns are no longer isolated from the domain of science as sources of possible contamination. A historically based image of science makes active citizen participation legitimate not only in the domain of technology but also in the domain of science. As a consequence, lay citizens could participate in targeting areas of preferred scientific research either directly by sitting on review panels or indirectly through their elected representatives.

CONCLUSION

We began this chapter by describing a historical episode from the 1930s. That episode embodies all the issues relevant to science and technology policy in the 1980s. When faced with the revelation that even the best scientists are fallible, the public was of two minds. On the one hand, it laughed to see the experts knocked down to the level of ordinary mortals, but, on the other hand, it was shaken to learn that its earlier faith in the omniscience of science might be misplaced. Thus began a period of public ambivalence about the social value of undirected scientific research. In the 1980s this ambivalence extends to technology as well. We would like to believe in the experts, but we are afraid that they do not have all the answers, particularly on questions of social welfare. In 1930 corporate leaders, among others, believed that science and technology required a new form of centralized, national management. But who should manage science and technology? Business executives? Scientists? Government bureaucrats? The public?

Answers to these questions require an analysis of many factors, all relevant to evaluating policy options for science and technology. A policy maker must consider the interactions of a policy with existing political institutions and with public values. He or she must consider the effects of past policy and of past scientific and technological inventions. But in addition, the policy maker must base policy on an adequate philosophical and historical understanding of science and technology and their interactions. In this chapter we have examined two images of science and technology from the perspectives of philosophy and history, and we have seen how these two images may direct policy in different directions. In our opinion, the results of historical and philosophical analyses support the new pragmatic philosophy of science and technology over the logical empiricist philosophy. Thus we believe that active public participation in setting guidelines for scientific research is justified. In the next chapters we will look at specific cases illustrating the interactions between science and technology policy, political institutions, and public values. We believe that these cases show that citizen participation in setting guidelines for science and technology is desirable as well as justifiable.

NOTES

1. The following account of the Millikan-Compton debate and its social context is documented by Daniel J. Kevles in *The Physicists: The History of a Scientific Community in Modern America* (New York: Knopf, 1978), chap. 16. Contemporary popular accounts of the debate are found in *New York Times,* September 15, 1932, p. 23; November 16, 1932, p. 16; December 3, 1932, p. 2; December 31, 1932, p. 1; January 1, 1933, p. 16; February 5, 1933, p. 1; and also in "Cosmic Row," *The Nation* 136 (January 18, 1933), p. 54.

2. Quoted in "The Revolt Against Science," *Christian Century* 51 (January 24, 1934), p. 110.

3. Frederick J. Schlink, "Government Bureaus for Private Profit," *The Nation* 133 (November 11, 1931), p. 508; D. W. McConnell, "The Bureau of Standards and the Ultimate Consumer," *Annals of the American Academy of Political and Social Science* 173 (May 1934), p. 150.

4. Frederick P. Keppel, "American Philanthropy and the Advancement of Learning," *School and Society* 40 (September 29, 1934), pp. 408–409.

5. Kevles, *op. cit.,* p. 249.

6. For a discussion of the ensuing policy debate, see Robert Kargon and Elizabeth Hodes, "Karl Compton, Isaiah Bowman, and the Politics of Science in the Great Depression," *Isis* 76 (1985), pp. 301–318.

7. This claim is supported by Robert Olby, *The Path to the Double Helix* (Seattle: University of Washington Press, 1974).

8. For examples of a logical empiricist philosophy of science see Percy W. Bridgeman, *The Way Things Are* (Cambridge: Harvard University Press, 1959); F.S.C. Northrop, "Einstein's Conception of Science," in *Albert Einstein: Philosopher-Scientist*, vol. 2, edited by Paul Arthur Schilpp (La Salle, Ill.: Open Court Publishing, 1949); and Karl Popper, *Conjectures and Refutations* (New York: Basic Books, 1962).

9. For an account of the role of truth within science, see Jacob Bronowski, *Science and Human Values* (New York: Harper and Row, 1965).

10. A tool-oriented conception of technology is discussed by Lewis Mumford, "Technics and the Nature of Man," in *Philosophy and Technology*, edited by Carl Mitcham and Robert Mackey (New York: Macmillan Free Press, 1972), pp. 77–85.

11. for a logical empiricist account of the scientific process, see Carl Hempel, *Philosophy of Natural Science* (Englewood Cliffs, N.J.: Prentice-Hall, 1966).

12. See James B. Conant, *Science and Common Sense* (New Haven: Yale University Press, 1967), p. 305; and James K. Feibleman, "Pure Science, Applied Science, and Technology: An Attempt at Definitions," in Mitcham and Mackey, eds., *op. cit.*, pp. 33–41.

13. Jon Elster, *Explaining Technical Change* (Cambridge: University of Cambridge Press, 1983). Elster discusses four major theories of technical change.

14. National Academy of Sciences, *Perspectives in Physics*, vol. 1 (Washington, D.C.: NAS, 1972), pp. 55, 61–62.

15. Robert J. Ackermann, *Data, Instruments, and Theory* (Princeton, N.J.: Princeton University Press, 1985), p. 36. See also Robert Merton, *The Sociology of Science: Theoretical and Empirical Investigations*, edited by Norman W. Storer (Chicago: University of Chicago Press, 1973).

16. *Ibid.*, p. 37.

17. Kargon and Hodes, *op. cit.*, p. 304.

18. Loren Graham, *Between Science and Values* (New York: Columbia University Press, 1981), pp. 217–256.

19. National Academy of Sciences, *op. cit.*, p. 399.

20. George Wise, "Science and Technology," *Osiris*, Second Series, 1 (1985), pp. 229–246.

21. *Ibid.*

22. Elster, *op. cit.*, pp. 145–146.

23. Kargon and Hodes, *op. cit.*, p. 302.

24. Wise, *op. cit.*

25. *Ibid.*, p. 233.

26. *Ibid.*

27. The following books are sources for researching the new philosophy of science: Harold I. Brown, *Perception, Theory, and Commitment* (Chicago: University of Chicago Press, 1977); Paul Feyerabend, *Against Method* (London: New Left Books, 1975); Russell N. Hanson, *Patterns of Discovery* (Cambridge: Cambridge University Press, 1958); Thomas Kuhn, *The Structure of Scientific*

Revolutions (Chicago: University of Chicago Press, 1970); and John Ziman, *An Introduction to Science Studies* (Cambridge: Cambridge University Press, 1984).

28. P. Rossi, *Philosophy, Technology, and the Arts in the Early Modern Era* (New York: Harper and Row, 1970), p. 149.

29. Peter Weingart, "The Relation Between Science and Technology—A Sociological Explanation," in *The Dynamics of Science and Technology*, edited by Wolfgang Krohn, Edwin Layton, Jr., and Pert Weingart (Dordrecht, Holland: D. Reidel, 1978), p. 260.

30. *Ibid.,* p. 267.

31. Wise, *op. cit.,* p. 240.

32. Weingart, *op. cit.,* p. 275.

33. Wise, *op. cit.,* p. 240.

34. Weingart, *op. cit.,* p. 274.

35. Wise, *op. cit.,* p. 242.

36. *Ibid.,* p. 235.

SELECTED READINGS

Barnes, Barry. *Scientific Knowledge and Sociological Theory*. London: Routledge and Kegan Paul, 1974.

Bloor, David. *Knowledge and Social Imagery*. London: Routledge and Kegan
Paul, 1976.

Brown, Harold I. *Perception, Theory, and Commitment: The New Philosophy
of Science*. Chicago: University of Chicago Press, 1977.

Ellul, Jacques. *The Technological Society*. New York: Knopf, 1964.

Graham, Loren R. *Between Science and Values*. New York: Columbia
University Press, 1981.

Kevles, Daniel J. *The Physicists: The History of a Scientific Community in
Modern America*. New York: Knopf, 1978.

Kuhn, Thomas. *The Structure of Scientific Revolutions,* 2nd ed. Chicago:
University of Chicago Press, 1970.

Mitcham, Carl, and Robert Mackey, eds. *Philosophy and Technology: Readings
in the Philosophical Problems of Technology*. New York: Free Press, 1964.

2
A DOUBLE WEDDING: SCIENCE AND GOVERNMENT, KNOWLEDGE AND POWER

For two hundred years the encounter between science and political power in the United States has had unique characteristics. Children of the Enlightenment and Newtonian mechanics, such as Washington and Jefferson, were proud of their scientific and technological achievements. Jefferson became well known in Europe and at home for his "natural philosophy" and architectural innovations. Revolutionary leader Thomas Paine designed the first suspension bridge and joined with Washington to conduct experiments on the Potomac to ascertain the cause of that river's mysterious "boiling" properties. In the process Washington, famed for tossing silver dollars across the Potomac, postponed the criticism that U.S. policy makers attempt to solve problems simply by throwing money at them.

The United States was born in the eighteenth-century Age of Reason and science; thus its leaders' interest in scientific endeavor is not surprising. Further, as the French philosopher Condorcet pointed out during the 1700s, "In every century Princes have been found to love the sciences and even to cultivate them, to attract Savants to their palaces and to reward by their favours and their amity men who afforded them a sure and constant refuge from world-weariness, a sort of disease to which supreme power seems particularly prone."[1] Although this patronage of science for its entertainment value is also part of the U.S. tradition, the earliest scientific activities of the U.S. government and citizens were primarily motivated by pragmatic applications. The French observer of life in the United States, Alexis de Tocqueville, said in the nineteenth century that "in America, the purely practical part of the sciences is admirably cultivated and great care is taken with the theoretical part immediately necessary to application; on this side the Americans display a spirit which is always clear, free, original and fertile, but there is practically no one in the United States who concerns himself with the essentially theoretical and abstract part of human knowledge."[2]

In the nineteenth century, English and French intellectuals viewed people like Benjamin Franklin and Paine as remarkable innovators and inventors rather than as theoretical scientists. In Tocqueville's eyes, the U.S. free enterprise system and its emphasis on applied research were responsible for this pragmatic approach to science. Without governmental interference or patronage, the inventive spirits of individuals like Franklin and Paine could roam freely over topics as varied as electricity and mechanical engineering. Political leaders like Washington and Jefferson, who were themselves interested in science, saw scientific endeavors in terms of private inventiveness and the marketing of new products, over which government should have no control. Thus, as presidents they set the early standard for political/scientific relations in the United States: Government would have little role in science other than to protect freedom of inquiry and to provide patents for new inventions.[3]

WORLD WAR II AND THE POLITICS OF SCIENCE

By most accounts, World War II was a watershed for the coming together of science and government in the United States. Before this time, presidents and the Congress rarely approached the scientific community for any reason; when they did, they exhibited the delicacy and wariness of explorers in uncharted lands. Notable (and exceptional) excursions into the realm of science included the signing of the Hatch Act in 1887 by President Grover Cleveland, which provided some public support of science through the establishment of agricultural research stations in each state, the founding of the National Academy of Sciences during the Civil War, and the 1898 appropriation by Congress, with some daring and much criticism, of $50 thousand to inventor Samuel Langley for the development of the "aeroplane." During the Great Depression President Franklin Roosevelt relied heavily on the scientists who dominated the newly established National Resources Board, though most were economists and social scientists, for advice on his economic programs.

Roosevelt also established and placed within the Executive Office the Science Advisory Board (SAB) to counsel him on technical matters; however, by 1940 the board's influence had faded and its recommendations to the president no longer found a sympathetic ear. The eclipse of the SAB can be dated at many points, though perhaps none as dramatic as October 11, 1939. On this date, President Roosevelt received a now famous letter from the United States' most renowned scientist, Albert Einstein, which read in part:

> Sir:
> Some recent work by E. Fermi and L. Szilard, which has been communicated to me in manuscript, leads me to expect that the element uranium may be turned into a new and important source of energy in the immediate future. . . . In the course of the last four months it has been made probable—through the work of Joliet in France as well as Fermi and Szilard in America—that it may become possible to set up nuclear chain reactions in a large mass of uranium, by which vast amounts of power and large quantities of radium-like elements would be generated.[4]

Thus began the first major joint effort in U.S. history between government and the scientific estate in which vast financial support was provided for both basic research and technical development of the atomic bomb.

The Manhattan Project that produced the bomb altered for the foreseeable future the relationship between science and government, but in a larger sense World War II itself, not the mechanism of its ending, accomplished this feat. Basic research carried out under public funds in the United States and elsewhere produced scientific marvels besides atomic weaponry, such as radar and proximity fuses for conventional bombs. Furthermore, the nature and uniqueness of this particular war had widespread effects not only on the relations between government and science but also on the nature of these two estates themselves. Neither government nor science in the United States came out of the war unchanged in each other's eyes or in the eyes of an increasingly watchful public.

From the U.S. standpoint World War II differed from all previous wars. Never before had all U.S. society been galvanized into a single, total war effort both in the field and in the factories and munitions plants at home. For most Americans it was and remains an effort of unequivocal moral uprightness.

The moral nature of the World War II experience greatly influenced the development of relations between science and government both during and after the war. During the war, the president assumed new responsibilities as the moral leader and keeper of morale in a society unified into a purposeful and crusading whole. These new presidential duties would be important for the scope of the U.S. presidency and for the development of U.S. science, since Roosevelt unilaterally decided to go forward with the Manhattan Project. Also, since the war was ended in part by the unleashing of a terrifying new technology, it is doubtful whether President Harry Truman could have ordered the bombing of Hiroshima and Nagasaki with the clear conscience that he professed if the war had not been perceived as the battle against a monstrous evil. Finally, because of the dropping of the atomic bomb, the war would effectively alter the way in which Americans and the rest of the world would think about future wars. After Hiroshima, few would agree any longer with the German historian and military strategist Karl von Clausewitz who stated that war is merely the continuation of diplomacy by other means. In the age of nuclear weapons, the next full-scale war might end not only diplomacy but civilization.

World War II did more than raise new fears about future wars; it fundamentally altered the relationship between science and government. The new relationship came about in part because the government brought the scientific community into the war effort in order to promote the development of nuclear power, radar, and new weapons. This general approach was not in itself new: Throughout history science has contributed mightily to myriad war efforts, from Archimedes' deadly geometry that predicted the trajectory of projectiles to DaVinci's innovative catapult designs. With World War II, however, two new developments emerged and took root. First, the close working relationship built up between the scientific and political communities during the war continued after its end. Second and more important for the long term, the administration and organization of the research effort during the war set the guidelines for the modern government/science relationship. As political scientist Don Price has observed, "the most significant discovery or development . . . was not the technical secrets that were involved in radar or the atomic bomb; it was the administrative system and set of operating policies that produced such technological feats."[5]

In its mobilization of science beginning in 1940, the U.S. government, specifically President Roosevelt, accepted a radically new assumption regarding the role of research during wartime. As Jean-Jacques Salomon pointed out, "Until then, military research had been content to adapt civil technologies to the needs of war, thus involving no radical innovation either in science or politics. During the second world war, scientific research was used for the first time as a source of new technologies whose influence was to be no less decisive on the post war period than on the termination of hostilities."[6] This new thinking concerning the military and political uses of scientific research necessitated a new administrative system for the public utilization of the products of that research. President Roosevelt responded by establishing two new wartime agencies, the National Defense Research Committee (NDRC) in 1940 and the Office of Scientific Research and Development (OSRD) in 1941. Together these agencies acted as administrative clearinghouses for federal grants and contracts that resulted in new military technologies. Many scientists were brought directly into government service through programs like the Manhattan Project; others remained in positions in university or government laboratories while working under federal contract.[7]

Critical to the success of the projects carried out under the auspices of the NDRC and OSRD was the degree of access to decision makers that the heads of these agencies would enjoy. In the case of the OSRD, its presidentially appointed head, Vannevar Bush, had direct access to the White House. Roosevelt had placed the agency in the Executive Office of the President and had made Bush his unofficial science adviser. This position enabled Bush "to deal with military leaders on equal or better than equal terms" and gave him "the leverage he needed in dealing with the vast network of administrative relationships on which the success of a Government agency depends."[8] The lofty positions granted Bush and OSRD during the war, and the vast resources placed at their disposal to complete projects like the development of the atomic bomb, brought science into politics and politics into science to a degree impossible to foresee before the war and impossible to reverse after the war.

POSTWAR HONEYMOON

The dramatic union of science and politics between 1940 and 1945 left an indelible mark on both estates after the war. Vannevar Bush's historic 1945 report to President Truman made it clear that future successes in any policy area depended upon the continuation of the symbiotic union between science and government forged during the war. In *Science: The Endless Frontier,* Bush declared that without "scientific progress no amount of achievement in other directions can insure our health, prosperity, and security as a nation in the modern world."[9] Therefore he urged that the federal government must begin to set policy for further scientific and technical development, for in doing so "lies much of our hope for the future."[10]

In the forty years since Bush's report, many issues of U.S. politics have involved scientific progress and science policy. This new preeminence for science—in Don Price's words, "the major establishment in the American political system"[11] has exacted a high price in terms of the self-image and public image of the scientific enterprise. Both inside and outside the scientific community, the stereotype of the scientist is no longer that of the lone researcher cloistered in a private laboratory. Science since the war has become a truly community activity: Competition for federal grants and private-sector investment is intense, and scientists find themselves answerable to new masters in government, industry, and society. In part, the communal nature of science is a legacy of vast cooperative efforts such as the Manhattan Project. But whatever the source, the image of modern science as a political establishment is far removed from Condorcet's view of the scientist as royal jester.

If the wartime experience liberated science from its role of merely providing diversion for those in power, it also inexorably levied upon it the burden of social utility. After Hiroshima, the idea that science's sole purpose was the pristine search for truth was largely exploded. The unleashing of atomic power proved in dramatic fashion that science could be put to use, and the continuation of the wartime relationship between science and government indicated that, at least in the eyes of government, the primary goal of research was social benefit. This change in the perception of science's goal would have far-reaching implications for both scientists and government officials. Scientists would now have to acknowledge and cope with the consequences of their actions. Government would have to recognize its own obligation to support scientific endeavor for the benefits that would thereby accrue to society as a whole.

These new perceptions of the nature and goals of science since World War II have been both the boon and the bane of the scientific community. On one hand, in comparison to earlier times scientists and engineers have found themselves virtually awash with funds with which to carry out research and development projects. Political leaders have consulted them, brought them into the halls of power, and hailed them as the high priests of modernity. On the other hand, as science has gained political clout it has come to be viewed by those in politics as an interest group, a source of power to be manipulated for political advantage. Salomon described the scientist's new situation as "circumscribed in an area of political decisions which affect his work and which are influenced by his work."[12] Science and government have formed a partnership in which each benefits and mutually exploits and manipulates the other. This union is an "alliance of science, in its specifically scientific character, with ideology as instruments in the hands of power."[13]

One result of the new relationship between science and political power is that government now has a major role in defining the objectives of scientific research. Government defines those objectives in purely instrumental terms: It orders the scientific community to provide ways and means of achieving goals that scientists do not set. Whether those goals are to reach the moon or to find a vaccine for the swine flu, the objective of scientific research is no longer knowledge for its own sake but for society's sake. And the objective is set by the agent charged with protecting the sake of society: government. Thus two hard lessons for scientists to learn after the heady rush of the Manhattan Project have been that they are not the philosopher-kings of the modern age and that informing political decisions is not the same thing as forming them.

SCIENCE AND POLICY FROM 1945 TO 1968: NSF

Since its genesis in 1950, the National Science Foundation (NSF) has been a bellwether of emerging policy relationships between science and government in the United States. As such it serves as a useful focal point for a discussion of the general outlines of science and technology policy over the past three decades. Of course, the NSF is only one of a myriad of science and technology agencies whose histories and mandates constitute U.S. science policy, and our focus on the NSF is not meant to belittle the importance of those agencies. Other relevant actors in the science policy process include agencies such as National Aeronautics and Space Administration (NASA), the National Institutes of Health (NIH), and the Defense Department; individuals or groups such as the president, Congress, interest groups, and citizens also play a large role in the formation of policy. But they do so largely because of the peculiar development of the NSF as a federal agency and because of its administrative style as set by its first director, Alan T. Waterman. Thus, the history of the birth of the NSF is a good place to begin our look at science policy making at the federal level.

Congressional Mandate of NSF

When President Truman signed Senate Bill 247, the National Science Foundation Act, into law on May 10, 1950, he performed a minor miracle, according to the author of the NSF's own official history.[14] The National Science Foundation Act was the last of twenty-one NSF bills presented in Congress since 1945, all direct progeny of Vannevar Bush's report on science to President Roosevelt and of the 1947 Report of the President's Scientific Research Board, entitled "Science and Public Policy."[15] Five years of debate in Congress and the press over the first twenty NSF bills made the final debate of historic moment and the NSF a creature of both pride and compromise.

By 1950 the idea of a civilian agency dedicated to the furtherance of basic research was no longer new. The wartime life of the Office of Scientific Research and Development had expired in 1945, but the resulting policy vacuum did not wait till 1950 to be filled. By the time NSF came into being, several other agencies had for several years been spending millions of dollars for basic research. Agencies in the Defense Department (DOD), most notably the Office of Naval Research (ONR), and new creations such as the Atomic Energy Commission (AEC) and the NIH were already in place and preparing the way for U.S. science policy. By 1950 the ONR alone was spending $40 million annually and subsidizing over 40 percent of the country's basic research under its director, Alan T. Waterman.[16] These agencies had continued the wartime practice of funding research largely through contracts rather than grants but had nevertheless moved federal support from the realm of military technology and applied research into that of basic research to a degree "the scientific community could hardly have imagined prior to 1940."[17]

The existence of the DOD, NIH, and AEC as sponsors of basic research meant that "NSF was born into a policy space filled with other research organizations."[18] As merely one agency among several charged with stimulating scientific growth, NSF could expect that it would be forced to compete for funds. Of course, NSF was the new kid on the bureaucratic block; thus it could also expect intrepid opposition both from Congress and from competing agencies for control of projects and money. This expectation was realized in its first operating budget as approved by Congress in 1951: NSF received an appropriation at the beneath-subsistence level of $225,000, in a national budget for basic research of well over $100 million.

The paltriness of the first NSF budget raised eyebrows and questions within the scientific community concerning the real level of support within Congress and the White House for an agency purportedly dedicated to the pursuit of scientific knowledge. In fact, Congress intended that NSF would be more than just the institutional substructure for the scientific ivory tower—making its budget decision more inexplicable. In the enabling legislation, Congress mandated that the new agency would not only support basic research; it also would be in charge of "correlating" and "evaluating" national science activities and would "develop and encourage the pursuit of a national policy for the promotion of basic research and education in the sciences."[19]

In short, Congress established NSF as the overseer of all other federal agencies, such as DOD and NIH, and expected it to be the progenitor of a complete science policy for the nation. Such expectations seem a bit ludicrous for a new agency within the nether world of bureaucratic politics and particularly for one that Congress had funded with such apparent reluctance.[20] Operating under an overreaching mandate but avoiding the role of overall policy maker, the NSF developed the present-day form for science and technology policy, and the history of NSF serves as a model for the development of that policy.

Initial Operation of NSF

When Alan Waterman resigned his post as director of the Office of Naval Research in 1950 to assume his duties as the first head of NSF, he accepted the post on one condition: that he be allowed to ignore much of the mandate Congress had given the new agency. Daniel Greenberg reported that "before taking the position [Waterman] exacted an assurance from the Bureau of the Budget that the fledgling foundation would not be required to fill a role that might bring it into conflict with far richer and politically more influential research agencies, such as the AEC and the Defense Department."[21] The Bureau of the Budget (BOB) would soon regret extending this assurance.

In Waterman's view, the assurance meant three things. First, it guaranteed that NSF would not be called upon to give either direction or official approval to scientific projects contemplated or currently under way in other agencies. Thus, NSF could reasonably hope to avoid ruffling bureaucratic feathers in other agencies, which might result in the loss of legislative support. Second, by not dictating to AEC and other agencies, NSF could avoid its roles as overall policy maker for science and as special adviser on science policy matters. Waterman felt that accepting such a policy or even advisory role would also force NSF to knock heads with other agencies and would lessen its scientific authority by tainting it with legislative or electoral, as well as bureaucratic, politics. The foundation would remain politically pure and therefore avoid making enemies that might endanger its primary function.

This function, as Waterman viewed it, was the third fruit of BOB's assurance. The foundation would concentrate its efforts on projects of pure, basic research and education and not on applied or technological projects. It would serve science, not society, and society would benefit by the increase of knowledge that resulted from NSF-supported research. However, how that knowledge would be put to use in society or how scientific gains could be implemented to better social life were concerns of other agencies. As budget officer William D. Carey reported to his superiors in 1952, "Waterman's idea is to operate on the basis of a scientist-to-scientist approach, instead of the agency-to-agency pattern. Out of this spiritual communion will emerge gentlemen's agreements leading to the formation of spheres of influence for research scholarship."[22]

Throughout his thirteen-year tenure as director of NSF, Waterman pursued what Lambright called a "science-oriented" administrative style. This approach meant adopting strategies that for the most part were self-limiting in terms of the authority NSF would wield and the amount of money it would dispense. As Lambright noted, Waterman adopted the ideology of science for the operating ideology of NSF. Scientific knowledge should be pursued for its own sake;[23] scientists should be granted maximum freedom in their funded research; minimum bureaucratic regulations regarding paperwork and official reportage should be required. As far as developing policy, Waterman proclaimed that science policy should emerge from scientists themselves and that in making such policy, any agency (including his own) "should defer to the judgement of the active and capable research scientists in the field."[24]

As NSF linked its fate to that of science and scientists, it ensured that its standing with the federal government and citizens would rise or fall with the image of science itself. For its first seventeen years, NSF pursued its self-defined purpose of supporting basic research almost exclusively without sustained criticism from either the Congress or the public. During those years the public accepted the federal role as the patron of science as necessary, partly because of a pervasive "endless frontier" ideology but more specifically because of the public perception of science's attachment to national security issues. Congress and the public defined the legacy of Hiroshima during the cold war of the 1950s and 1960s as the U.S. drive to remain ahead of the Soviet Union in both basic science and technology. This objective meant enormous federal expenditures for weapons research throughout the 1950s, and, with the surprise launch of Sputnik in 1958, a new technological race in space research.

These developments do not appear to be immediately relevant to NSF's history during its first two decades. In a sense this is true, for weapons and space research were clearly the big budget items in science and technology expenditures during this period, and Congress placed them largely under the purview of the Defense Department agencies and NASA. However, during the 1950s and 1960s the political environment surrounding scientific research was a benevolent one.

Congress and the president approved vast sums of federal money to support basic science research, relaxed accounting procedures for the use of funds, and took few steps to counter the decentralized nature of the science support apparatus within the federal government. Science was a top priority in federal spending, and NSF flourished in the reflected glow of astonishing scientific advance.

Increased governmental support for science did not mean, however, that NSF played an increasingly large role in science spending or policy in the years before 1968. In fact, quite the opposite is true. Though NSF's budget increased dramatically, its share of federal research and development (R&D) money fell equally dramatically throughout the period when compared to that for other agencies such as the Defense Department, NASA, or NIH, which were also supporting primarily basic research projects. NSF was growing in its ability to serve science but at a far slower rate than that of other agencies. Furthermore, because NSF through its science-oriented operating ideology continued to evade its mandated role in setting national science policy, its political position vis-à-vis other agencies became increasingly precarious. NSF occupied a rarified realm of erudite scientists engaged in arcane scientific research, and it rarely entered the realm of politics and had little political clout. By 1969 the political mood had changed regarding the value of pure research and the need for federal budget cuts.

SCIENCE POLICY AFTER APOLLO

"If we can put a man on the moon, why can't we . . . ?" This plaint which echoed across the countryside and appeared on newspaper opinion pages in late 1969, evinced that the historic Apollo moon landing signaled a new attitude toward science and its social function in the United States. The lunar landing was an unparalleled achievement in scientific discovery and technological innovation. Furthermore, as an overflow from the space race, exciting new technologies and innovations became available for social use and everyday life, such as an array of new plastics and petroleum polymers. NASA was masterful both in pursuing a dream of scientific research and in announcing the resulting useful new products. Thus the agency successfully justified to the U.S. people the vast sums necessary to reach the moon on other grounds than merely the "search for knowledge." However, once the landing was accomplished, NASA and the scientific community quickly discovered that neither the entertainment value nor the incidental innovative gadgetry would be enough to ensure widespread governmental or public support.

Decline in Public Support for Science

The reasons behind this shift in public attitudes toward science are complicated and difficult to assess. Certainly in the late 1960s an amorphous fear developed toward scientific discoveries that arrived so rapidly that public digestion and understanding lagged far behind.[25] Also, in this time of "flower children" and dropouts, a vague feeling that technology had exceeded the ability of democratic institutions to control it was a pervasive theme in the popular media and arts. Perhaps the U.S. public had simply become jaded by the enormity of scientific advance in the previous two decades. Certainly the dramatically lower figures for the television audience watching the event of the second lunar landing compared to the first would bear out the decrease in interest.

Of course, the late 1960s and early 1970s also were a period of vast social and political upheavals. The war in Vietnam and the crises in U.S. cities carried implications not only for traditional politics but also for the politics of science and science policy. Domestic issues such as poverty, civil rights, and urban problems were not, as Lambright stated, "R&D intensive," and even the Vietnam war was fought with traditional rather than developing technology.[26] Furthermore, inflation caused by the war effort naturally cut into federal research budgets. Finally, as protests against the war erupted on college and university campuses across the country, higher education institutions fell out of governmental favor as seats of scientific research conducted with federal funds. Without the universities, scientific research as a national priority was clearly waning. Lambright concluded that "in every sense this was a period of economic recession for R&D. It was one of transition in public/governmental attitudes toward science and technology. It left many scientists, engineers, industry and university executives, and even government officials bewildered and embittered."[27]

The decline of public and governmental support for science from 1969 to approximately 1974 had more distinctly political causes as well. The tremendous achievement of the Apollo program was possible largely because of an initial and continuing presidential belief in the importance of the project. After the lunar landing, it became rapidly evident that no other science project currently under way or under consideration enjoyed such a presidential mandate. As a result, without an external stimulus comparable to the Soviet launch of Sputnik in 1957, President Richard M. Nixon decided that a politically possible, even desirable, approach would be to lessen the federal government's role in scientific research.

The political factors behind Nixon's decision are not difficult to recognize. First and foremost, coordination within the scientific establishment in Washington was lacking for either policy setting or scientific advice. The reason for this lack lay with Alan Waterman's decision to ignore NSF's congressionally mandated role in setting science policy. NSF quite simply was not a presence in the Nixon White House. This became important in Nixon's decision to cut back on scientific research funds even though NSF had played a small role in the now-completed triumph of Apollo. Because NSF by its own (or rather its director's) choice exerted little political clout, cutting the budget for the type of program it represented—pure research— was an easy political decision.[28]

Nixon did not stop with cutting funds for many R&D projects. In a move of far greater significance and potential damage for the entire scientific community, the president razed the presidential advisory system for science and technology. In reorganizing the Executive Office in early 1973, Nixon abolished the Office of Science and Technology, established in 1962, terminated the White House post of science adviser to the president, informally established by President Roosevelt, and asked for the resignations of the members of the President's Science Advisory Committee (PSAC). In addition, Nixon moved the presidential science advisory function to the National Science Foundation and the National Security Council. Neither of these groups was notably equipped to deal with its new task, particularly NSF because of its history of avoiding the job of giving such advice. Thus, as David Z. Beckler, a member of the PSAC, lamented, "in one fell swoop, the President eliminated the entire White House science and technology mechanism that had been painstakingly erected in the years following the Soviet Sputnik in 1957."[29]

Specific political events involving scientific consultants and the entire scientific community spurred the president to these administrative actions. Congressional debates concerning the antiballistic missile (ABM) program and the supersonic transport (SST) project included testimony by experts who doubted the efficacy and cost-effectiveness of both ABM and SST. The president had used the full power of his office to support these programs; thus their subsequent defeat caused much friction between the White House staff and the resident scientists. In addition, this period was characterized by an increase in political activity by scientists throughout the country. Prominent scientists, many with experience on now-defunct presidential advisory organs, spoke out on political issues like the ABM and lent their names and credentials to opposing candidates for the presidency in the next election. The Nixon White House did not respond warmly to this new political awareness among scientists; indeed, the president reacted by further cutting funds for pure research projects.[30]

Of course, not all actions regarding science and technology during the Nixon administration involved cutting back federal support for science and technology. New programs initiated by the administration concerned energy research and the fight against cancer. In both cases the administration took innovative steps and appropriated considerable funds in these new research areas. Both initiatives, however, were the results of legislation first proposed by Congress in response to widespread favorable public sentiment and public interest group lobbying. Congress responded to Nixon's elimination of the science advisory system by establishing its own group in 1972—the Office of Technology Assessment. The objectives of this new agency were to grant additional leverage to Congress in science and technology affairs and to aid it in exercising control over science agencies by requiring them to consider the long-range implications of their research and development projects.

REVIVING SCIENCE: FORD, CARTER, AND REAGAN

From the perspective of the scientific community, the Nixon administration's attitude toward scientific R&D and its practitioners could charitably be described as ambivalent. However, toward the end of the brief Ford administration the stage was set to alter once again the course of government/science relations. In his last month in office, President Gerald Ford sent to Congress a report entitled *Issues '78,* which was attached to his final budget message. Well received by Congress, this document redefined the need for strong federal government support for scientific research and development, and it formed a basis for science policy in the Carter and Reagan administrations.

In *Issues '78* Ford argued for a continuance of increased federal support of scientific research begun in the Ford administration's 1977 budget but went further by providing a rationale for long-term R&D funding. The federal government had for two decades accepted a responsibility for funding basic research, the report stated, and its grounds for doing so—national interest—not only had not changed but had become more pressing. The national interest in increasing federal support for science was evident because basic research was crucial for the well-being of U.S. society in two areas: (1) the economy and (2) health, the environment, and energy.[31] The federal role must be expanded, the report continued, because private industry could not be expected to pay for costly research whose profitability would only be seen in the long term. However, in areas of research in which "the technical risk is low, development time is relatively short, or an intimate knowledge of the market is required," responsibility for R&D should be left to private industry.[32]

Thus the Ford administration sought to reverse the trend away from widespread federal support for scientific R&D so apparent in the Nixon years. This policy also meant that NSF enjoyed renewed patronage by the Ford administration. According to *Issues '78*, for fiscal year 1978 Ford proposed a budget increase of 13 percent for NSF and argued that the agency had a role in funding applied research, principally through the Research Applied to National Needs (RANN) program. In addition, Ford charged NSF with improving science education in the United States. Science education activities received $76 million in the proposed 1978 budget, including funds to improve equipment in school laboratories and for programs to attract women and minorities to careers in science.

Renewed Commitment

The Ford administration took symbolic measures intended to emphasize its renewed commitment to scientific R&D. President Ford reestablished the Office of Science and Technology in the Executive Office and appointed a science adviser to head the newly recreated President's Science Advisory Council. On all fronts science was making a comeback in the White House, and perhaps as a result of definitive presidential leadership, the Congress and the country warmed once again to the promise of scientific and technological advancement.

When Jimmy Carter took up occupancy in the White House in 1977, the office of the president was for the first time in fifty years filled by a man with professional training in the sciences. As a nuclear engineer, President Carter claimed special insight into the nation's needs in the area of scientific R&D. Carter, who continued in the spirit of *Issues '78*, also provided his own initiatives that would increase the federal role in supporting R&D. Carter supplied a somewhat new rationale for federal support of science; he claimed in his 1979 State of the Union address that such support constituted not merely federal patronage but investment.

> Scientific research and development is an investment in the nation's future, essential for all fields, from health, agriculture, and environment to energy, space, and defense. We are enhancing the search for the causes of disease; we are undertaking research to anticipate and prevent

significant environmental hazards; we are increasing research in astronomy; we will maintain our leadership in space science; and we are pushing back the frontiers in basic research for energy, defense, and other critical national needs.[33]

Although the new Carter approach to science funding was largely a rhetorical and symbolic shift of emphasis, it nevertheless signified "an important revision and advance in the attitude of government officials toward public expenditures for science and technology, particularly basic research."[34] So strongly did Carter feel the need for federal support that he personally took the lead in persuading Congress to protect basic research appropriations. He wrote Congress during the budget committee hearings in 1978 that

> I want to emphasize that even relatively small reductions in key agencies—such as the National Science Foundation—or in new initiatives and growth planned for the mission agencies—including NASA and the Departments of Agriculture, Energy and Defense—would defeat our objective. Modest increases in real growth in these programs are necessary if we are to strengthen the nation's capacity and productivity in critical areas of research.[35]

In fact, during the Carter years federal spending for basic research rose a total of 14 percent, with NSF enjoying one of the largest gains of over 50 percent.

Much of the increase in research spending during the Carter administration centered on energy research, such as that for development of synfuels and other alternative energy sources, and on the space shuttle—a program continuing from the Ford years. Although specific policy moves of the administration between 1977 and 1981 will be dealt with in the next six chapters, one aspect of the Carter science policy is worth noting for its relevance to assessing science and technology decisions in the Reagan administration. Between fiscal years 1978 and 1982, research spending increased dramatically, particularly in areas under the aegis of the Department of Defense (DOD). In fact, DOD spending for research rose over 100 percent—from $324 million to $714 million—catapulting DOD from fifth to third place among agencies receiving basic research appropriations.[36] This trend was enthusiastically embraced by policy makers in the Reagan White House.

Reagan's Mixed Record

The record of research and development support in the first years of the Reagan administration can best be described as mixed. Overall federal spending for scientific R&D increased through 1983, but marked shifts in the institutional location of R&D funds and in the rationale behind federal support for science became increasingly evident. Presidential Science Advisor George Keyworth signaled these changes in a December 1981 statement to the House Committee on Science and Technology: "The federal role in science and technology must be different from that which has prevailed since World War II. It must be appropriate to the 1980s—appropriate to a national mood which calls for increased vigor and acceptance of responsibility by individuals and organizations in the private sector and decreased involvement by the federal government in many of our affairs."[37]

For the Reagan administration the term *appropriateness* meant primarily that an increased amount of the R&D money spent in this country must come not from government but from the private sector. Though President Reagan and his policy advisers acted at least symbolically according to the spirit of *Issues '78* by invoking its name in administration budget reports and by publicly acknowledging Carter's view of research appropriations as investment, they also sought to fit R&D policy into their general approach to economics and federal spending. First, they appropriated an increasing part of basic research funds to the Department of Defense. In some cases they transferred funds from other agencies directly to DOD. For example, the administration decimated the NSF fund for graduate research support and social science research, while increasing the appropriations to DOD for precisely those purposes.

Second, the economic theory guiding the administration moved President Reagan to rely increasingly on tax credits to corporations to stimulate research and innovation in the private sector by other means than government spending. Though Reagan's approach did not include some of the more radical proposals put forward by staunch fiscal conservatives—such as Milton Friedman's recommendation to withdraw support entirely from basic research and abolish the National Science Foundation—the administration did cut back considerably in its funding for civilian basic research. Thus, for example, NSF suffered a loss in appropriations in real dollars of 2.3 percent.

The Reagan approach to promoting new research by means of tax credits to private industry had some drawbacks. Administration critics argued that this approach did not indicate much sensitivity to or understanding of the way in which U.S. industry views expenditures for research leading to innovation. Lewis M. Branscomb, chairman of the National Science Board and vice president of International Business Machines (IBM), charged in 1981 that the administration's use of an incremental R&D tax credit to spur innovation would have little impact precisely because industry did not view R&D in the same manner as did public agencies: "The fact is that technology-based corporations like IBM have been steadily increasing their R&D investment over the past years, and because the tax credit is only for *incremental* increases, IBM and other corporations like it probably won't get much tax advantage out of it because we are already spending what we consider about the right amount on R&D"[38] (emphasis in original).

Although it can be argued that certain aspects of the Reagan approach to scientific R&D were fundamentally flawed, in other respects the administration was merely recognizing a trend within the emerging science policy process in the United States. Since at least 1968, that trend has been away from an exclusively governmental responsibility for science policy making and toward involvement in the policy process by other, nongovernmental, actors. The Reagan administration partially acknowledged this development by recognizing through the tax credits program the role of industry in setting science policy goals. However, the administration's recognition of new actors in the field of science policy did not go far enough; corporations are not the only possible nongovernmental actors in the policy process.

CONCLUSION

In this chapter we have looked at the history of science policy in the United States primarily as a chronicle of developing governmental attitudes toward research and development and of the agencies that grew up and embodied these attitudes. But others besides agencies and elected officials contribute to science and technology policy. As the Reagan administration implicitly acknowledged through its tax credit program, corporations also make decisions about the type of technologies to develop and the kinds of scientific research avenues to pursue. Furthermore, citizen groups, special and public interest groups, lobbying groups, and individual citizens also have a large and increasing role to play in the development of science and technology policy, particularly when policy must be made amid widespread public controversy. In the following chapters, we will closely examine a number of such controversial policy areas and seek to elucidate the ever more complicated and populated arena in which science and technology policy is formulated.

NOTES

1. Quoted in Jean Jacques Salomon, *Science and Politics* (Cambridge, Mass.: MIT Press, 1973), p. 3.

2. Alexis de Tocqueville, *Democracy in America,* vol. 2, edited by J. P. Mayer (Paris: Gallimard, 1951), p. 14.

3. Salomon, *op. cit.,* p. 25.

4. Quoted in *The Atomic Age,* edited by M. Grodzins and E. Rabinowitch (New York: Simon and Schuster, 1965), p. 11.

5. Cited in U.S. Congress, House Committee on Science and Astronautics, Subcommittee on Science, Research, and Development, *Report: Toward a Science Policy for the United States* (Washington, D.C.: Government Printing Office, 1970), p. 81.

6. Salomon, *op. cit.,* p. 48.

7. W. Henry Lambright, *Governing Science and Technology* (New York: Oxford University Press, 1976), p. 16. See also Don K. Price's discussion of the four varieties of government contracts and forms of administering them in *Government and Science* (New York: New York University Press, 1954), pp. 68–72.

8. Price, *op. cit.,* p. 45.

9. Vannevar Bush, *Science: The Endless Frontier* (Washington, D.C.: National Science Foundation, 1960), p. 11.

10. *Ibid.,* p. 12.

11. Don K. Price, "The Scientific Establishment," *Proceeding of the American Philosophical Society* 106 (June 1962), pp. 1099–1106. Quoted in Joseph Haberer, *Politics and the Community of Science* (New York: Van Norstrand Reinhold, 1969), p. 186.

12. Salomon, *op. cit.,* p. xx.

13. *Ibid.*

14. Milton Lomask, *A Minor Miracle: An Informal History of the National Science Foundation* (Washington, D.C.: National Science Foundation, 1976).

15. Cited in Dorothy Schaffter, *The National Science Foundation* (New York: Praeger, 1969).

16. Lomask, *op. cit.,* p. 72.

17. Lambright, *op. cit.,* p. 18.

18. *Ibid.,* p. 142.

19. *National Science Foundation Act,* Public Law 507, Section 3, 81st Congress.

20. The nature of bureaucratic politics in the formation of science and technology is elucidated by Lambright, *op. cit.,* and will be discussed in subsequent chapters.

21. Daniel S. Greenberg, *The Politics of Pure Science* (New York: New American Library, 1967), p. 144.

22. Lomask, *op. cit.,* p. 93.

23. Lambright, *op. cit.,* p. 145 passim.

24. U.S. Congress Report, *NSF: Its First Fifteen Years,* 1966. Quoted in Lambright, *op. cit.,* p. 146.

25. See Robert Blank, *The Political Implications of Human Genetic Technology* (Boulder, Colo.: Westview, 1980).

26. Lambright, *op. cit.,* p. 21.

27. *Ibid.,* p. 23.
28. David Z. Beckler, "The Precarious Life of Science in the White House," *Daedalus* 103 (summer 1974).
29. *Ibid.,* p. 115.
30. *Ibid.,* p. 124.
31. Claude E. Barfield, *Science Policy from Ford to Reagan* (Washington, D.C.: Atomic Energy Institute, 1982), p. 3.
32. Quoted in Barfield, *ibid.,* p. 5.
33. *Ibid.,* p. 12.
34. *Ibid.*
35. *Ibid.,* p. 14.
36. *Ibid.,* p. 17.
37. *Ibid.,* p. 40.
38. Quoted in Barfield, *ibid.,* p. 60.

SELECTED READINGS

Barfield, Claude E. *Science Policy from Ford to Reagan.* Washington, D.C.: Atomic Energy Institute, 1982.

Brooks, Harvey. *The Government of Science.* Cambridge, Mass.: MIT Press, 1968.

Bush, Vannevar. *Science: The Endless Frontier.* Washington, D.C.: National Science Foundation, 1960.

Conant, James B. *Modern Science and Modern Man.* New York: Doubleday, 1952.

Ellul, Jacques. *The Technological Society.* London: Cape, 1965.

Greenberg, Daniel S. *The Politics of Pure Science.* New York: New American Library, 1969.

Kuhn, Thomas J., and Alan L. Porter, eds. *Science, Technology, and National Policy.* Ithaca, N.Y.: Cornell University Press, 1981.

Lambright, W. Henry. *Governing Science and Technology.* New York: Oxford University Press, 1976.

Lomask, Milton. *A Minor Miracle: An Informal History of the National Science Foundation.* Washington, D.C.: National Science Foundation, 1976.

Schaffter, Dorothy. *The National Science Foundation.* New York: Praeger, 1969.

Salomon, Jean-Jacques. *Science and Politics.* Cambridge, Mass.: MIT Press, 1973.

3

ENERGY: PROBLEMS
OF SOCIAL CHOICE

For most Americans energy did not become a political issue until the 1970s. The provision of energy was not a matter of difficult or even social choices. True, each individual home owner or company manager had to decide between coal, oil, natural gas, or perhaps nuclear-generated electricity to heat his or her establishment, but this choice was personal and private. Since the first Arab oil embargo of 1973 this situation has changed, of course, and energy questions and energy policy constitute much of the current and future political agenda. Nevertheless, a discussion of this change must be mostly speculative: Even though the need for an effective federal energy policy has been clear since 1978, the sacrifices that such a policy would demand on the part of citizens have made its implementation difficult.

Case studies abound that seek to highlight the energy dilemma in which the United States—and much of the developed world—presently finds itself. Perhaps none of these portrays the radical change in the public mind concerning energy as well as the countless personal sagas about waiting for hours in long lines of idling automobiles hoping to receive a ration of a suddenly scarce commodity: gasoline. This tableau of frustration was played out in 1978 and has been repeated in several isolated instances since that time. The current energy crisis is made up of countless similar personal experiences.

During the first Arab oil embargo in 1973 and shortly after, U.S. energy problems were commonly characterized as foreign-policy difficulties. According to political commentators and politicians, arrogant and truculent Arab sheiks had temporarily interrupted the flow of oil to the West, but such was said to be the nature of diplomacy in an unsettled part of the political world. Order—and gasoline—would be restored when those who were to blame resolved their political disputes. In short, the United States was not to blame for its sudden energy shortage, and soon all would be resolved and the flow from a seemingly unlimited supply of oil and natural gas would be restored.

Except for some well-informed citizens and government officials, few Americans even speculated that the country's energy shortage might be caused not by a suddenly powerful foreign oil cartel but by years of profligate energy consumption by U.S. citizens. Since the end of World War II, U.S. oil and gas usage had risen at an annual rate of 3.4 percent; in the years between 1965 and 1973, it rose at a rate of 4.5 percent.[1] This increasing demand for oil and natural gas made the United States heavily dependent on foreign oil, a large portion of which came from Gulf countries. As a result, the energy-dependent economy of the United States was increasingly vulnerable to sudden price variations or interruptions in the supply of Middle Eastern oil, as the gasoline lines of 1973 made depressingly clear.

U.S. allies who were also major importers of Middle Eastern oil were affected by the 1973 embargo far more than was the United States. Japan was almost totally dependent on foreign oil, and Germany and France were not far behind in their reliance on the goodwill of Arab oil exporters. The message of 1973 should have been abundantly clear: The energy problems of the West were the result and responsibility of Western economic policies that did little to encourage either conservation or the development of alternative energy sources.

In the United States between 1973 and 1978, oil imports doubled while prices rose because of inflation. Public opinion polls during the period indicated that most Americans felt that potential energy shortages had more to do with Arab greed than with U.S. shortsightedness in reliance on foreign energy sources. Not until the fall of the shah of Iran in 1979 and the resultant skyrocketing of Iranian oil prices under Ayatollah Khomeini did the U.S. public rouse itself into accepting conservation measures and begin to clamor for a coherent energy policy. However, in the eyes of many, the time for conservation and policy making on a grand scale had already passed. Conservation was no longer enough, and policy that dealt only with energy questions could not by itself ensure a future exempt from catastrophic energy shortages. Dorothy Zinberg argued that

> Had the government mounted a concerted conservation effort in 1973, after the OPEC oil embargo, oil imports would likely have decreased. Instead, by 1979 imports had almost doubled, while prices had tripled. . . . In the short space of a decade, the United States has striven to develop a comprehensive energy policy to subsume all of

the concerns identified above. [We] contend that the aim of achieving such comprehensiveness was misguided and unrealistic. Because the issues are ultimately political and the country so diverse, a more realistic approach is to explore the assumptions and the values underlying policy planning and choices.[2]

In this chapter we will see that the issues of energy extend beyond the technical questions surrounding various new technologies and beyond the vagaries of foreign policy. What Americans have learned from the 1970s is that energy questions are not merely personal choices or concerns but involve all U.S. society in making perplexing social decisions. These decisions affect all citizens and therefore give rise to policy disputes about who should make energy policy and what the role of citizen participation should be in the policy process. Furthermore, the problems of energy provision raise ethical and constitutional questions that transcend the boundaries of energy technology and reach to the very fiber of U.S. political ideology.

ENERGY, ETHICS, AND POLITICS

In a presidential address in 1977 Jimmy Carter announced that the energy crisis in the United States constituted the "moral equivalent of war." Though this statement was largely received as a piece of political grandstanding, President Carter was pointing to the new energy environment that OPEC had wrought. As discussed in Chapter 2, during wartime presidents take on a special symbolic role, and the citizenry looks up to them for inspiration amid the widespread sacrifices that the war effort requires. In his energy speech, Carter may have invoked the metaphor of war to signal to U.S. citizens that the future provision of energy would entail the same type of personal sacrifices. This message was important because the energy problems that the United States continues to face will require more than technological fixes that demand no sacrifice by an energy complacent public. Furthermore, energy policy can no longer focus entirely on the technological side of energy provision; it must also take into account the ethical issues present in all policies that demand sacrifice.

Energy Policy as an Ethical Issue

The reasons for an ethical approach are not immediately apparent. The United States is an energy-rich nation, and alternative energy technologies are already in place to be utilized if foreign oil supplies are unavailable in the future. Why then can we as a nation not rely on these technologies and the fixes they offer to our energy woes?

The answer lies in the fact that each new technology, as well as new policy approach to oil reliance, embodies difficult ethical questions, and the responses to these queries involve almost impossible political choices by policy makers at all levels of government.

Many potential solutions to U.S. energy problems exist, but, as economist Lester C. Thurow pointed out, all of them pose extremely difficult problems of social choice and political decision making.[3] These solutions essentially reduce to two general courses: continue reliance on oil as an energy source, but alter the pricing and distribution system; or wean the United States from its dependence on oil and natural gas as its primary sources of energy. The second alternative will be examined later in this chapter and in Chapter 4; the first choice involves unique problems that illustrate the peculiar ethical and political problems of energy in the United States.

Supply and Demand Pricing

Since 1981 the energy policy of the administration of Ronald Reagan has consisted of an increased reliance on the private sector. In its most comprehensive statement concerning energy, *The National Energy Policy Plan,* published in July 1981, the administration described its energy policy as an example of its overall political philosophy regarding the role of government and the private sector: "The administration's reformulation of policies affecting energy is part of the president's comprehensive program for economic recovery, which includes elimination of excessive federal spending and taxes, regulatory relief, and a sound monetary policy. The aim of the economic recovery program is to release the strength of the private sector."[4] As this policy related specifically to reliance on oil as a primary energy source, it advocated a supply and demand pricing system, which, the administration assumed, would both decrease energy consumption and spur investment in alternative energy sources.

Reliance on the marketplace to set the price for petroleum has an obvious appeal for several reasons. First, this policy embodies traditional free enterprise values and is therefore particularly alluring to the Reagan administration. Second, using supply and demand pricing for oil would presumably raise the price for oil to a level that would allow increased profits to be used for investment and exploration for new oil wells. Third, as classical capitalist theory concludes, a high price for oil would stimulate competition in the industry. High profits would then lead to undercutting of prices as new companies enter the field, and ultimately the consumers would reap the benefits.

However, this general policy of relying on oil to meet energy needs has two serious problems when applied to the existing market situation. The first results from the artificial nature of the actual world oil market; the second from the very real nature of the redistribution of income such a policy would require. According to capitalist theory, high prices for a commodity should cause new investors to enter the field, forcing prices down as competition with older, established companies becomes acute. However, as Thurow pointed out, these exertions of the market system occur if the price of a commodity is set by nature—by the costs imposed in the production of a product. If a near-monopoly cartel like the Organization of Petroleum Exporting Countries (OPEC) sets prices, none of the rules of the market apply. A cartel that has manipulated prices upward has the economic strength to undercut the price of any potential competitor without suffering a major economic loss.

> The price of imported oil is high because of a cartel, not because of natural scarcities. . . . Since the price is a man-made price, it becomes very difficult to make alternative energy investments. . . . Suppose that imported oil is selling for $30 per barrel. Now imagine that you were able to discover some process that would make synthetic oil for $25 per barrel. Could you afford to go into production? If the $30 price were set by Mother Nature, the answer is a clear "yes." . . . If the $30 price is set by a cartel, however, going into production is not so clear-cut. Massive investments would be necessary to make the synthetic oil; but what would happen once you started producing oil? The cartel would simply cut their price and you would be left with a large worthless production facility.[5]

U.S. oil companies have realized the complexities that Thurow identified in a supply and demand pricing system for oil, and the dangers of investment of which Thurow speaks apply to oil exploration for new wells as well as to new synthetic oil-producing processes. Consequently, since the deregulation of oil and natural gas prices began in 1978, U.S. oil companies have not been in a race to plow new profits back into the industry in the shape of exploration or discovery of synthetics. Instead, the uncertainty of a cartel-dominated oil market together with high prices has led to the frenzy of mergers and corporate takeovers that has characterized the U.S. economy since the second oil embargo in 1978. In other words, the high profits realized by the industry have led to diversification by major oil companies into unrelated markets, rather than to the widespread discovery and exploitation of new oil wells. And though the new

Exxon copiers and microcomputers may be admirable products, they do little to fill the gasoline and oil tanks of U.S. consumers.

The second difficulty with supply and demand oil pricing as a way to resolve U.S. energy problems has less to do with economics than it does with ethics and traditional political problems surrounding the distribution of wealth in a society. If market pricing were somehow to overcome the manipulation of OPEC, the resultant high prices for oil products would cause a massive redistribution of income among U.S. consumers. As consumers pay a higher price for energy, a larger percentage of their income would go for this commodity, and great shifts in the distribution of wealth would take place. Because energy is a precious commodity, such redistribution involves issues of social or distributive justice—the fairness of the apportionment of benefits or wealth in a society. Redistribution would be controversial under any circumstances—the oil companies become richer and the consumer poorer—but even among consumers the distribution of increased costs would not be equitable.

Because most Americans mainly rely on sources of energy like gasoline, higher oil prices would affect all citizens. However, citizens in certain regions of the country (e.g., the Northeast, which relies heavily on oil to meet its energy needs) and poorer citizens generally would suffer a larger proportional drop in their incomes if the price of oil were fully deregulated. As Thurow pointed out, "the proportion of income going to energy consumption differs dramatically between the rich and the poor and less dramatically across regions. . . . The real income effects among the poor are almost seven times as large as they are among the rich."[6]

Distributive Justice

Inequities in the distribution of wealth cause tremendous political problems for U.S. policy makers because they fear that citizens would perceive the distribution as unfair. And though dilemmas of social justice plague policy makers in any society, they are particularly acute in the United States, where concepts of distributive justice have been formulated at least since the New Deal as involving the distribution of advantages or benefits. In the context of deregulation of oil prices, however, the focus of distribution is on costs or sacrifices; and this type of negative focus is persuasive, as President Carter seemed to have understood, only during wartime and analogous situations.

The problem of equity in the distribution of costs and benefits plagues other actual or potential energy technologies as well. For instance, U.S. coal supplies and reserves are abundant enough to provide energy well into the twenty-first century at the very least, but advanced energy technologies utilizing coal require many difficult policy choices. Transforming coal into natural gas is a technology already in place and upon which energy experts place much hope. However, the highest grade coal for this process is found primarily in western states such as Montana and Wyoming, where mining is expensive and the vast quantities of water needed for the gasification process are not available. Who is willing to give up their water in the arid West in order to provide energy for the rest of the country? Probably no one. The only other option for gasification is to move the coal to the northeastern states where water is abundant. Transporting the coal would be a viable, though expensive, alternative. However, another problem arises in this approach: The processing of coal into natural gas would result in mountains of residual slag and tons of airborne fly ash. Whose land is going to be used for the necessary slag heaps? is the obvious question; shipping the slag back West would render the natural gas derived from gasification far more expensive than oil.[7] Already in the Northeast hundreds of lakes are dead or dying because of acid rain, the result of fly ash spewed into the atmosphere by Midwestern coal-fired utilities and industries. The obvious conclusion from this example is discouraging: The costs of using coal as a primary energy source may be low only as long as environmental concerns are ignored.

Problems concerning the equitable distribution of energy costs and benefits have resulted in what Kevin Phillips calls the political phenomenon of "the Balkanization of America."[8] States and regions of the country are squabbling over energy like Eastern nations argue over Balkan warm-water ports. Acid rain, for example, has motivated regional confrontations and even lawsuits between states. The political truth illustrated here is that with respect to energy, the interests of some conflict with interests of others. Only two alternatives have become evident in these clashes of self-interest: cooperation with sacrifice or stalemate. Stalemate has prevailed in the United States since energy and the environment became political issues in the 1970s.

The political and ethical issues generated by the use of coal as a primary energy source surround most other energy technologies as well. Nuclear power, for example, once seemed like a panacea for cheap energy. But since 1978, the doldrums have descended upon this industry, as witnessed by the many closed and permanently unfinished nuclear power plants dotting the countryside. The optimistic projections of the past were based on an overly simplified picture of the political and social realities of nuclear energy. Put simply, nuclear power is cheap as long as one ignores the risks of irradiation; however, once risks enter the picture, the costs rise. The distribution of risk is itself a perplexing social choice (investigated fully in Chapters 3 and 4), but risk as part of both the cost/benefit equation and the pattern of distribution is a reality not yet fully acknowledged by policy makers.

Alternative Technologies

Finally, a host of so-called alternative technologies presents similar political and ethical difficulties concerning the distribution of costs and benefits. Power generated from solar, wind, synthetic fuel, geo-thermal, and biomass conversion sources could act as an inexpensive energy source for specific tasks in some areas of the country but not in others. Because of their limited applicability, these technologies do not represent large-scale alternatives in the United States' energy future.

At the moment solar energy seems to many to be the most attractive alternative energy source. Solar technology has obvious advantages such as its lack of dangerous waste materials and its inexhaustible energy source. Also, the generating machinery it utilizes has few, if any, moving parts and can operate virtually unattended with little maintenance. Furthermore, a photovoltaic (PV; solar energy) system is silent and has little environmental impact other than the placement of the solar collectors.[9]

Until recently the primary drawback to solar energy was its impracticality for large-scale energy needs. The sheer number and size of the requisite solar collectors (panels) for any large energy project seemed to lessen the technology's potential. Also, after an initial period of intense governmental support for solar projects under President Carter, the interest at policy levels has waned considerably, and the Reagan administration has cut funding for solar energy research by over 50 percent per year. Still, since 1970 federal agencies have spent over $700 million on solar PV systems, and most observers believe that spending from the private sector has exceeded that amount.[10]

Private-sector investment in solar energy systems at both the householder and corporate levels was spurred by the passage of the Public Utility Regulatory Policies Act of 1978. Under this act small producers (those using less than 80 megawatts) of electrical energy were allowed to sell excess electricity back to utility companies, which were forced to purchase it on the principle of "avoided cost." This cost is "the cost of energy to the utility which, but for the purchase, the utility would generate itself or purchase from another source."[11] Utility companies began to purchase privately produced electricity in significant amounts in 1980, and the utilities' interest in generating solar power themselves began to increase accordingly. In fact, the Electric Power Research Institute (EPRI) reported that since 1976 the number of reported solar PV projects undertaken by utility companies has increased from five to seventy-four, and the number of utilities reporting projects has risen from five to forty-one.[12] This prospect is encouraging for solar energy advocates, and with new developments in the storage and collection of solar energy, it promises renewed hope for the United States' energy future.

The very high capital costs of solar conversion at the utility level remain, however, and raise yet again the difficult problem of the distribution of costs. Because utilities are under public control, cost inequities that arise cause political problems as utilities pass on any additional costs to consumers. Reliance on the private sector for energy innovation has resulted in an increase in interest in solar energy and in considerable investment by private citizens and corporations. This reliance has been in keeping with the Reagan administration's strategy for energy policy but does nothing to resolve the long-term policy requirements for energy provision for the United States. Even with the new progress in solar energy, we are essentially back where we started this chapter: with plenty of technological answers to the question of energy but with few political answers to the policy and ethical questions raised by the new technological innovations.

DOE AND AN EQUITABLE POLICY FOR ENERGY

In 1980, the Department of Energy (DOE) completed a study entitled "Energy Programs/Energy Markets" that attempted to assess the long-term effects of the major energy programs undertaken by federal agencies since the first Arab oil embargo in 1973. In the study, the DOE used economic models to predict the expected impact of these programs in the 1990s as compared to the hypothetical energy picture for that time if the programs had never been initiated. Harvey Brooks reported that the study's conclusion was surprising: "The conclusion is startling: the net effect of all federal energy programs is almost precisely zero. Total United States energy consumption will be about 2 percent less in 1990 than it would have been in the absence of these programs, while imports will be about 200,000 barrels per day, perhaps 3 percent, higher."[13] Such speculation is somewhat questionable, of course, given unforeseen technological and political developments. Yet this conclusion indicates two things: first, that the Carter and Reagan policies of energy price deregulation most probably will have a more pronounced conservation effect than federal energy programs; second, that price decontrol will likely exacerbate the income inequities among Americans, as energy price increases will continue to hurt most those who can afford them least.

Ad Hoc Energy Policy

These foreseeable consequences of an energy policy of deregulation and price decontrol point to a common failing of much U.S. science and technology policy: a failure to come to grips with the economic and social impact of ad hoc policies generated to respond to temporary political necessities. If a policy of deregulation and decontrol would lead to inequitably distributed and adverse economic effects in the population, these consequences would in the long run spell the policy's political failure. In the long run, politics, particularly electoral politics, would guide the generation and persistence of policies aimed at the solution of U.S. energy problems. If a large (or sometimes even a small but vocal) minority of U.S. voters persistently feel that their interests are being ignored in matters of policy, eventually they will articulate these interests in a national political forum in such a way as to call the relevant policies into question. An important point to recognize here, as James Madison did in 1789 in Federalist Paper #10, is that actual discrimination is not necessary as long as the perception of discrimination exists and is strong enough to convince a minority that its voice is not being heard and its interests are not being considered. When such a perception or a reality exists over time, electoral pressures would be applied to change policies sufficiently to alleviate discrimination.

This encapsulation of U.S. electoral and policy politics may seem overly simple, even idealistic, to members of a group subject to real or imagined discrimination. However, this view of politics and policy making guides U.S. politicians, as evidenced in President Carter's "moral equivalent of war" speech. President Carter seemed to make two appeals in this speech. First, he acknowledged that sacrifice in the energy crisis would be necessary but would hurt some groups more than others. Knowing this, he appealed to the affected groups for a compromise that would achieve results beneficial to the whole of society first and to their own interests second. Second, he asked that sacrifice of personal or group interest be accepted in the patriotic spirit that Americans manifest in times of crisis.

These appeals are necessary, Don Price argued, because most often the problems faced by government are not those caused by inefficiency or administrative ineffectiveness. Rather, they are caused by the inability to achieve compromises among the conflicting goals, values, and interests of competing groups in U.S. politics.[14] U.S. policy problems then require two things in Price's view for their resolution: first, an adequate and sometimes new institutional context for the airing of concerns and interests; and second, institutional measures and political symbols capable of convincing reluctant citizens to accept sacrifice and of rewarding them for doing so. In the history of the energy crisis, U.S. institutions have failed to meet either requirement, despite several attempts.

Founding of the DOE

Perhaps the first and most far reaching institutional change made to deal with the energy crisis was the establishment of the Department of Energy by Congress and President Carter in 1977. President Nixon originally proposed the idea of a separate cabinet-level department charged with overall authority in energy-related matters as part of a general bureaucratic reorganization plan centering in the Department of the Interior. Lost amid the turmoil of Watergate, the plan was resuscitated by Gerald Ford and finally approved in somewhat altered form by Congress in mid-1977. After the predictable congressional wrangles and compromises aimed at limiting the power of the new agency, James Schlesinger became the first energy secretary in October.

The DOE absorbed most of the energy-related functions of the federal government, including in entirety those of the Energy Research and Development Administration (ERDA). ERDA had only been in existence since 1974, when Congress established it along with the Nuclear Regulatory Commission (NRC) to take over the duties of the old Atomic Energy Commission. The DOE was to have wide-ranging authority over much of the federal energy bureaucracy, but its enabling legislation did little to avoid potential political conflicts between DOE and other energy-related agencies. For instance, the old Federal Power Commission, which had regulatory authority over natural gas companies and prices, remained an independent agency, though subsumed within the DOE organization.[15] In refusing to place all authority for energy matters in the hands of DOE, Congress ensured the continuation of fragmented policy making and interagency power struggles that had plagued the politics of energy since 1945.

Though the DOE would not be an institutional remedy for the political and policy quandaries of energy, its establishment served an important symbolic function. As argued privately within the Carter administration, "the creation of a Department of Energy could be a potent political symbol that something concrete was being done to address the much-discussed shortages of secure supplies of energy in usable forms."[16] Such symbols of government action are important aspects of presidential and bureaucratic politics, but unless they are given sufficient mandate to make widespread policy changes, they are doomed to lose their symbolic meaning and become merely another forum for interest group politics.

Subsystem Politics

In fact, the DOE has suffered this fate. Because energy has become a highly charged political issue, new and competing constituencies or interested groups have come to realize their stake in an adequate energy policy. This realization is largely the result of the distributional aspects of energy cost that affect the income of citizens in disparate, even inequitable, ways. These constituencies or interest groups make their voice heard in Washington through lobbying and pressure politics. Typically, such politics takes the form of what Lawrence Dodd and Richard Schott called "subsystem triangles," consisting of the staff of a specialized congressional committee or subcommittee, a bureau or office in an executive agency, and relevant lobbying groups or corporations.[17] Though subsystem politics is pervasive and necessary in U.S. politics, it is often unproductive in providing widely accepted policy on emotion-laden issues like energy. This approach is often ineffective because large portions of the U.S. public not represented by lobbyists or subsystem triangles see the policy resulting from subsystem politics as narrowly ideological and even unfair.

A primary reason for the perception of inequity that has plagued energy policy is the nature and workings of the Department of Energy since its inception. As Harvey Brooks elaborated: "Nowhere have the workings of subsystem politics been more apparent than in the operations of the Department of Energy. With its mission pieced together through countless separate and independent legislative initiatives, the department has struggled to be all things to all people and to appease hundreds of different public constituencies."[18]

The DOE in its few years of operation has become a forum for every special interest group concerned with energy, each pushing its own favorite and surefire solution to the ongoing energy crisis. Providing such a forum was in fact one of the symbolic purposes of the DOE when it was introduced by President Carter, but to fulfill this function the agency required either legislative or executive leadership to overcome the paralyzing effects of subsystem politics. Leadership has never been forthcoming, though much legislation and agency policy have wound up in the courts. Without leadership from Congress or the president, the department has floundered amid a host of conflicts involving divergent policy goals, and "the mission of DOE [has become] the accretion of hundreds of overlapping and sometimes conflicting statutory provisions and policy philosophies."[19] Examples of the conflicting policy goals confronted by the DOE were listed by Brooks:

> 1. the rate of expansion or substitution of energy supplies versus the level of protection of public health and the environment; 2. the desirability of energy conservation versus public resistance to higher prices or detailed government regulation of consumer choices and behavior; 3. holding down domestic oil and gas prices to avoid economic dislocation and hardship for particular groups and regions versus the hazard of further stimulating dependence on Persian Gulf oil imports through the *de facto* subsidization of such imports through price controls and entitlements; 4. responding to public resentment against internal wealth transfer to domestic oil producers and multinational oil companies versus export of U.S. wealth to OPEC; 5. avoidance of risks to public health and the environment versus energy security and the diversification of U.S. energy supply sources; 6. reducing the risks of nuclear weapons proliferation versus the maintenance of good relations with our allies and our friends in the developing world; 7. reliance on the market to determine investments in energy supply and conservation versus the use of public policy and government regulation or targeted subsidies for this purpose; 8. our obligation to shape our domestic energy policies to make scarce and expensive oil available to assist economic development in the LDC's versus our concern for our own economic health and the protection of our own environment and safety.[20]

Compromise is difficult, if not impossible, on so many interconnected conflicts because of the perception of inequity or unfairness that accompanies them. Thus, it is no wonder that DOE has never become the panacea for U.S. energy woes that it was predicted to be. Responsibility for this failure must be placed squarely on the shoulders of members of Congress and particularly on the president, for as Jimmy Carter understood, these conflicts represent crisis, and in a crisis so analogous to war, citizens have the tendency—and right—to look to their president for direction. If the president is unwilling or unable to respond, the subsystems take over.

ENERGY POLICY, LEADERSHIP, AND THE PUBLIC

The failure of DOE to establish a coherent and successful energy policy was not the result of poor management but of a lack of leadership, both political and moral. For leadership Americans do not look either to bureaucratic departments or to the cabinet secretaries of those departments. They look to their elected representatives in the Congress and White House. If leadership is not forthcoming from those places, citizens in a democracy press their own narrow interests either themselves or through interest groups. Thus subsystem politics results. But, as Jimmy Carter realized in 1977, and as observers of U.S. energy policy and DOE have realized since, subsystem politics is not the solution to the United States' problematical energy future.

Within a democracy compromise is the approach most readily adopted in the face of difficult policy choices. However, the parties to compromise in subsystem triangles are motivated by divergent but private self-interests. Instead, national compromise requires concessions of private interests in deference to an overall national or public interest. Keeping the national interest paramount requires sacrifice among all subsystem actors, indeed among all citizens. The key here is to demand an equitably distributed sacrifice for the public interest. Such equity cannot be found, or even sought, in subsystem politics in which self-interest reigns and some interests are not even represented. Only in the realm of public interest can all private interests be considered equally, even if the resulting amounts of sacrifice are themselves different. In other words, only in the name of the public interest can all citizens be asked to sacrifice their self-interest, for only there can all believe that their interests have been recognized. This trust is a matter of faith, of course, but the hallmark of the U.S. presidency is that, because he or she alone is elected by the country as a whole, only the president represents all Americans and thus embodies the public interest itself.

Therefore, the president must take the lead in energy policy. He must convince U.S. citizens that sacrifice is necessary, that such sacrifice will probably be more severe for some than for others, and that all sacrifices will be acknowledged. The first two messages are already apparent to most Americans. The third is crucial to avoid the cynical descent into subsystem politics. Compromises are essential that keep "the national interest paramount while affording some, though by no means complete, protection to the various special interests affected."[21] Equal protection of interests does not necessarily mean equal treatment but equal *consideration* of all affected interests. The president, as the elected representative of the public interest, ultimately provides this guarantee of equal protection.

How do presidents or the country as a whole manifest this guarantee of equal consideration? During the crisis of war this is done by compensating those who have sacrificed most. Some compensation is symbolic, for example, medals, promotions, some of which are awarded posthumously. Some compensation is very tangible, like that provided by the GI Bill. Compensation is crucial because it is an acknowledgment that sacrifice has been made—and appreciated. Compensation should also be available for individuals or groups that sacrifice their personal welfare or interest in the name of national energy self-sufficiency. Both symbolic and tangible compensation are conceivable, ranging from conservation awards for individuals to tax cuts or increased public works funds for those communities that contribute most to the public interest in energy provision. These programs do not involve radical policy departures for either the president or the country as a whole, for they have been common during wartime crises. If President Carter was correct that the energy crisis is indeed the moral equivalent to a wartime crisis, then perhaps it should be responded to in a similar way. This response can be accomplished not by the establishment of new forums for subsystem politics but by moral leadership and the recognition of sacrifice for the public interest.

NOTES

1. Dorothy S. Zinberg, ed., *Uncertain Power* (New York: Pergamon, 1983), p. xxiv.

2. *Ibid.,* pp. xxiv, xxv.

3. Lester C. Thurow, *The Zero-Sum Society* (New York: Basic Books, 1980), p. 28, passim.

4. Department of Energy, *The National Energy Policy Plan,* report to the Congress required by Title VIII of the Department of Energy Organization Act (Washington, D.C.: DOE, July 1981), p. 3.

5. Thurow, *op. cit.,* p. 39.

6. *Ibid.,* p. 29.

7. *Ibid.,* p. 37.

8. Kevin Phillips, "The Balkanization of America," *Harper's,* May 1978.

9. Edgar A. DeMeo and Roger W. Taylor, "Solar Photovoltaic Power Systems: an Electric Utility R&D Perspective," *Science* 224, no. 4646 (April 20, 1984), p. 245.

10. *Ibid.*

11. Public Law 95-617, Public Utility Regulatory Policies Act of 1978, Section 210 (d).

12. DeMeo and Taylor, *op. cit.,* p. 245.

13. Harvey Brooks, "History as a Guide to the Future," in Zinberg, *op. cit.,* p. 219.

14. Don K. Price, *America's Unwritten Constitution* (Baton Rouge, La.: Louisiana State University Press, forthcoming). Cited in Harvey Brooks, *op. cit.*

15. Donald R. Whitnah, *Government Agencies* (Westport, Conn.: Greenwood Press, 1983), p. 113.

16. *Ibid.*

17. Lawrence C. Dodd and Richard L. Schott, *Congress and the Administrative State* (New York: Wiley, 1979.) These alliances are sometimes referred to as "iron triangles" or even as "unholy trinities" in the political science literature.

18. Brooks, *op. cit.*, p. 228.

19. *Ibid.*

20. *Ibid.*, pp. 228–229.

21. *Ibid.*, p. 231.

SELECTED READINGS

Burton, Dudley J. *The Governance of Energy.* New York: Praeger, 1980.

Casper, Barry M., and Paul David Wellstone. *Powerline: The First Battle of America's Energy War.* Amherst, Mass.: University of Massachusetts Press, 1981.

Illich, Ivan. *Energy and Equity.* New York: Harper and Row, 1974.

Kendall, Henry W., and Steven J. Nadis, eds. *Energy Strategies.* Cambridge, Mass.: Ballinger, 1980.

Landsberg, Hans H., et al. *Energy: The Next Twenty Years.* Cambridge, Mass.: Ballinger, 1979.

Rosenbaum, Walter A. *Energy Politics and Public Policy.* Washington, D.C.: Congressional Quarterly Press, 1981.

Schmidt, F. H., and D. Bodansky. *The Energy Controversy.* San Francisco: Albion, 1976.

Stobaugh, Robert, and Daniel Yergin, eds. *Energy Future: Report of the Energy Project at the Harvard Business School.* New York: Random House, 1979.

Zinberg, Dorothy S. *Uncertain Power.* New York: Pergamon, 1983.

4
NUCLEAR POWER AND SOCIAL JUSTICE

In our discussion of energy policy in Chapter 3, we saw indications of a growing level of public involvement in energy issues, which would increase the difficulty of policy making. In part citizens are pressing their claims for participation because of a redefinition of issues on the political agenda: Even though many policy questions may appear to rely largely on scientific or technical factors, the public often does not view them that way. Instead, science and technology policy issues are recast in the political arena as ethical questions that involve equity, justice across generations, individual rights, or personal freedom. On such questions, the U.S. political heritage admits of no experts, for the doctrine of liberal individualism has always been fully grounded in the ideas of individual moral independence and responsibility. Thus, reliance on expert opinion (and the presumed deference by citizens that this reliance implies) often is not enough to arrive at an effective and successful policy decision.

In no policy area is the phenomenon of issue redefinition more apparent than in siting policy—that is, decisions made concerning the construction and placement of public-works projects like airports, dams, or power plants. Particularly in the siting of nuclear power plants, the reluctance of the public to defer to expert scientific advice is growing and becoming an effective hindrance to the development of nuclear technology. Similarly, many of the arguments against the construction of nuclear generating plants rest on value judgments and definitions of justice: matters concerning which, according to the California state legislature, "the voter is no less equipped to make judgements than the most brilliant Nobel laureate."[1]

Citizen involvement in issues surrounding nuclear technology has brought new actors into the policy-making process. This broadening of the cast of characters, which is not new in democratic politics in the United States and which indeed is common in other science policy areas, signals a widespread challenge to traditional policy making concerning scientific and technological development. Allowing private citizens a powerful voice in technical decisions can frequently hinder the decision making process for those charged with conducting it, as we will see in the Midland case study. However, denying citizens this participation is politically risky, since it endangers democratic political institutions. As political theorist C. Wright Mills once observed, the stability of democracies is threatened when citizens feel that "they live in a time of big decisions; they know they are not making any."[2]

NUCLEAR POWER IN MIDLAND

On December 14, 1967, Consumers Power of Michigan announced its intention to begin construction on a new nuclear generating plant in the small town of Midland, located approximately 200 miles northwest of Detroit. Citing a growing consumer demand for electricity, the utility planned to share construction costs with the nearby Dow Chemical Company, which would receive 4 million pounds of process steam per hour from the plant as a byproduct from the electricity generation process.[3] The plant was scheduled to be completed by 1975, the utility announced, at a cost of $349 million. Seventeen years later (on July 16, 1984), the utility's board of directors at Midland voted to halt construction on the plant because of rising costs, decreasing electrical demand, and public concerns over safety. At the time of termination, the project was 85 percent completed and had cost nearly $3.5 billion.

Organized Opposition and Legal Channels

The controversy that led to the demise of the Midland plant is a textbook example of the problems besetting the nuclear industry. Though the history of nuclear power in Midland presents some unique characteristics, it has been repeated in numerous siting disputes throughout the United States. These disputes invariably manifest a number of similar elements: an array of vociferous antinuclear groups of both national and local origins, a changing regulatory environment, lengthy legal battles, and judicial victories for the nuclear industry that nevertheless result in the failure of its construction plans.

In Midland, the announcement of the site stirred the formation of a group of local citizens called the Saginaw Valley Nuclear Study

Group, headed by Mary Sinclair. The group's initial purpose, according to Sinclair, was not to oppose construction but to raise questions aimed at ensuring the operational and environmental safety of the plant. At the initial licensing hearings before the Atomic Safety and Licensing Board, which issued a provisional construction permit (the ASLB was part of the Atomic Energy Commission), the small study group of thirty members joined with other national antinuclear groups and submitted a list of nine thousand safety-related questions to be answered before a construction permit should be granted. Included in the coalition of six groups opposed to construction without safety assurances was the United Auto Workers, which provided considerable financial support for the licensing battle. The public hearings for consideration of a construction permit lasted for more than two years, during which period construction was halted.

In December 1972, the AEC approved the Midland construction permit, and site preparation resumed. However, during the lengthy licensing hearings, the U.S. Supreme Court handed down its decision in the *Calvert Cliffs Operating Committee* v. *AEC* case. The court ruled that all proposed construction permits for nuclear power plants must be accompanied by an environmental impact statement as required by the National Environment Policy Act, passed by Congress in 1969. This ruling delayed the Midland permit for several months, but more important, it enabled the groups opposed to the plant to pursue their opposition in the federal appeals court after the ASLB had ruled.

According to the Atomic Energy Act of 1954 and the Administrative Procedure Act of 1946, the issuance of construction permits for nuclear power plants may be appealed to the U.S. Courts of Appeals "after all possible administrative remedies have been exhausted."[4] After the ASLB had issued the permit for Midland, administrative remedies were in fact exhausted, so Mary Sinclair's group took the case to the District of Columbia Circuit Court of Appeals. The choice to lodge the application in the District of Columbia circuit court rather than in the circuit court for Michigan because of the former's historic sympathy toward antinuclear intervenors was apparently a wise one: In the decision passed down on July 21, 1976, the circuit court agreed with several points made by the Saginaw intervenors in the permit hearing of 1970–1972. As a result the court remanded the Midland construction permit back to the licensing board of the Nuclear Regulatory Commission (NRC) (which had replaced the old AEC in 1974) and ordered construction at Midland halted until a new permit could be issued. The court also ordered the NRC to reevaluate its procedures in issuing permits, stressing that the agency must exhibit a greater concern for full public disclosure of investigation results, for the problems of the disposal of radioactive waste, and for energy conservation in future licensing deliberations.

So construction on the Midland plant was delayed again, but the NRC appealed the case to the Supreme Court, which allowed construction to continue during the duration of the case. After hearing verbal arguments from Consumers Power and the Saginaw intervenors in November 1977, the court issued its unanimous decision in favor of the utility on April 3, 1978. Writing for the court, Justice William H. Rehnquist strongly rebuked the actions and decision of the appeals court, accusing it of exhibiting "judicial intervention run riot."[5] Stating that for more than four decades the Supreme Court had sought to restrain itself and lower courts from interfering with the authority and procedures of regulatory agencies, the court ordered that the original license for the Midland plant be reinstated and construction be allowed to continue to its completion.

Changing Economic and Political Climate

After almost eleven years, the utility had won its right to construct the nuclear plant at Midland. However, much had changed since its original decision to build the reactor. The old AEC had been replaced by the Nuclear Regulatory Commission, which had issued hundreds of new regulations to be followed during construction. The price of construction had risen astronomically because of inflation, legal fees, and the costs of complying with the NRC's new safety regulations. Also, demand for electricity had dropped dramatically from 1967 estimates, as the spirit of conservation had taken hold of the country. Thus the need for the new plant, so apparent in 1967, was now suspect.

Finally, the political environment of nuclear power had evolved into one of outright antagonism toward the further development of the technology. The public fears and the contentions of antinuclear groups had been given dramatic new credence by the accident at the Three Mile Island reactor in Pennsylvania and by the eerie timeliness of the movie *The China Syndrome,* which described fictionally a potential nuclear disaster only two weeks before the Three Mile Island incident actually occurred. In the face of strong opposition, the fact that nuclear-generated electricity no longer even cost less than that provided by coal-fired plants seemed to take away all reasons for proceeding with further construction. Thus, the Midland board decided to halt construction of the nearly finished plant.

The events leading to the cancellation of the Midland plant are by now part of a familiar and often-repeated scenario within the nuclear industry. Since the late 1970s over one hundred orders for new nuclear plants have been canceled, eighteen in 1982 alone. Though eighty-two nuclear plants are now operating in the United States with few prospects for premature and permanent shutdown, the forty-eight additional plants in various stages of completion face numerous political, regulatory, and financial hurdles. Together these obstacles make the likelihood of any unfinished plant ever generating a kilowatt of electricity very doubtful.[6] What had begun as a technology of such great promise in the 1950s was now, as *Time* magazine trumpeted on a 1984 cover, apparently "bombing out."[7]

DEVELOPMENT OF U.S. NUCLEAR POLICY

Though perhaps premature, the eulogies said over the nuclear industry in the early 1980s were a far cry from the optimism that greeted the technology's introduction in 1946. By the end of the Manhattan Project in 1945, the U.S. government had spent approximately $2 billion on the creation of a new industry. Scientists who had been involved in the project that had won the war looked with hope to the development of peaceful uses for the atom. The military, however, sought to retain control of atomic development, and in 1946 a bill was introduced in Congress that represented this point of view. The May-Johnson bill proposed the establishment within the Department of Defense (DOD) of an Atomic Energy Commission that would take charge of the development of nuclear projects. Scientists lobbied hard against the bill, arguing that atomic weapons must remain under civilian control. The scientists' viewpoint prevailed, and later in the same year the first Atomic Energy Act was passed establishing the AEC as an independent agency charged with encouraging and regulating all atomic energy projects.[8]

Early Military-Civilian Program

Though the first legislative act regarding atomic energy specifically reduced the DOD to an advisory role, the military legacy of atomic power would affect for forty years its development for civilian use in the United States. The arms race for nuclear superiority that raged between the United States and the Soviet Union in the 1940s and 1950s dictated the type of civilian nuclear power that would be developed in the United States. In the United States civilian nuclear technology has been built upon military nuclear technology because the type of civilian reactors constructed and employed in the country is that preferred for military reasons. U.S. electricity-generating reactors are all the water-cooled type and use enriched uranium for fuel. Enriched-uranium plants were decided upon because this fuel was already being manufactured for nuclear weapons. Thus, the enriching plants that already existed for military purposes could also be employed for a peaceful purpose—the provision of fuel for civilian nuclear power.

Enriched-uranium reactors are far more dangerous than those in Europe that use natural uranium for fuel. Because the heat generated by the degeneration of enriched uranium is many times that of natural uranium, U.S. nuclear plants are more efficient than their European counterparts, but they are also far more susceptible to breakdown and even possible meltdown.[9] Because the risk of meltdown gives rise to a major fear among U.S. nuclear power opponents, the initial choice in the 1940s to develop one type of nuclear technology has had a major impact on the controversy that has stymied the growth of the industry in the 1980s.

In the early years of the nuclear industry, however, optimism reigned. By 1950 the atomic bomb program had become the country's largest industrial enterprise, and by 1953 the AEC actually "owned and operated three towns, employed 5% of the nation's labor force, and consumed 10% of the country's electric power."[10] Civilian uses for nuclear technology were also being explored. In spring 1953 a small government reactor in a national laboratory in Idaho generated electricity for the first time, and in December President Dwight D. Eisenhower's famous "Atoms for Peace" address to the United Nations began a rush toward the civilian and peaceful exploitation of the new technology.

Encouragement Coupled with Regulation

The second Atomic Energy Act of 1954, which followed Eisenhower's speech, built upon the enthusiasm then displayed within the nuclear industry. To encourage private industry to exploit the technology, the act directed the AEC to subsidize companies willing to construct nuclear generators and to purchase for military purposes the plutonium that was a byproduct of the use of enriched uranium as fuel. In support of the belief that "the goal of atomic power at competitive prices will be reached more quickly if private enterprise, using private funds, is . . . encouraged to play a far larger role in the development of atomic power,"[11] the act directed the AEC to promote the construction of nuclear power plants by electric utility companies and gave the agency the power to issue licenses for construction and operation of new plants.

Thus, in the 1950s federal policy for civilian nuclear power focused on the encouragement of the technology for peaceful uses. The AEC's Power Demonstration Reactor Program (PDRP) in the mid-1950s was the flagship in this effort and offered to utility companies a variety of incentives to experiment with different designs for nuclear plants.[12] As stated previously, the light-water, enriched uranium model eventually became the design of choice for the AEC, and by the end of the decade the commission announced its readiness to consider applications for construction permits.

The second Atomic Energy Act also gave the AEC regulatory power over the industry. This approach put the agency in the peculiar position of promoting the idea of nuclear conversion to private utility companies, while seeking to control that conversion through its power over licensing and safety procedures. This arrangement was also unusual in that most U.S. industries are subject to regulation only after they have been established. Because of the recognized dangers of the new technology, however, the AEC's regulatory power had to be flexed side by side with its power to encourage nuclear development. Furthermore, the AEC exercised regulatory authority during the entire life cycle of each reactor from its construction to its eventual decommissioning.

> Regulation meant establishing standards by which the safety of plants was judged, commission review of designs and operating plans against those standards, and inspection of the construction and operation to ensure that all conditions were met. Each utility that planned to construct a power reactor had to apply to the commission for a construction permit; once the reactor was constructed, the utility needed a license to operate the reactor. During reactor operation, the utility had to comply with relevant commission regulations.[13]

Given this unusual degree of regulation regarding construction and operation, utility companies in the late 1950s were slow in jumping aboard the nuclear bandwagon, despite the enticements of the Power Demonstration Reactor Program (PDRP).

Profitability Versus Risk

Notwithstanding the financial incentives offered to utility companies through the PDRP, officials of every major corporation with potential interest in nuclear power doubted the profitability of nuclear generation of electricity. Citing the potentially catastrophic effects of a major breakdown at a plant, officials argued that the legal expenses incurred following an accident would "financially destroy a company."[14] To allay such fears, Congress in 1957 passed an amendment to the second Atomic Energy Act that limited the financial liability of a utility in case of accident. Named the Price-Anderson Act, it guaranteed that corporate liability for any damage caused by an accident at a plant would not exceed $560 million. Initially the guarantees were only to be in effect for ten years, but they have since been extended twice by Congress and remain in effect until 1987.

Without the financial security granted by the Price-Anderson Act, it is doubtful that the U.S. nuclear industry would have experienced the tremendous growth during the 1960s and early 1970s. But even with the guarantees, the margin of profit for nuclear plants remained very small. Therefore, companies pressed the AEC to approve rapidly construction and operation permits, since every delay represented a substantial economic loss. Throughout the 1960s the AEC attempted to develop regulatory criteria that would enable it to review applications quickly and uniformly. However, as the number of applications grew with accelerating speed and the power levels of proposed plants increased eightfold to tenfold to achieve economies of scale, the AEC found itself unable to keep up. Consequently, the power companies found that they were unable to make money.

By the early 1970s utility companies had found nuclear power an increasingly risky financial proposition. Delays in licensing had inflated costs and postponed, if not eradicated, profits. Furthermore, the National Environmental Policy Act (NEPA) of 1969 brought the nuclear industry under the additional regulatory purview of the Environmental Protection Agency (EPA). And in 1971, the *Calvert Cliffs* case imposed on the industry the responsibility of providing expensive environmental impact statements with every licensing application to the AEC. The *Calvert Cliffs* decision also required the AEC to broaden its regulatory standards with respect to environmental impact concerns relating to licensing; thus application procedures were slowed even further.

By 1972 considerable grumbling was heard in the nuclear industry over the AEC's inability to perform one of its two mandated functions in the nuclear energy field—the encouragement of nuclear development. However, public concern over the safety of nuclear power had also increased, and the dual encouragement/regulatory functions of the AEC led to charges of governmental bias in favor of nuclear power and against safety. Thus in 1974 Congress passed the Energy Reorganization Act, which abolished the AEC and replaced it with two new agencies. The Energy Research and Development Administration (ERDA) would take over research and development in all fields of energy, including nuclear. All regulatory functions concerning nuclear power would now reside in the Nuclear Regulatory Commission (NRC).

The Bottom Falls Out

This division of the old AEC's functions between separate agencies stilled some criticism of the government's handling of nuclear power, but throughout the 1970s public pressure and opposition to nuclear power continued to increase. Even after the first oil embargo in 1973 resulted in higher prices for nonnuclear fuels, the nuclear industry saw its own costs soar in part because of increased public opposition. Public hearings over license applications, similar to those in Midland, became increasingly lengthy, and legal costs mounted as NRC decisions were appealed in the federal court system. As nuclear power became a persistent political issue, the industry consistently found it less of a profit-making proposition.

By the mid-1970s utility companies were canceling by the dozens orders for new nuclear plants. This trend continued throughout the decade, and when the public's fears of a nuclear accident were made concrete in 1979 at Three Mile Island, even the ongoing construction of new plants ground to a halt at several sites. Much has been written about what actually occurred at Three Mile Island and about the extent to which the safety of the public was actually in jeopardy.[15] The accident obviously was a serious one, and it was compounded by a combination of mechanical and human errors. Although debate continues over the lessons either the nuclear industry or the NRC should have learned from this event, in retrospect one fact is indisputable: At Three Mile Island the bottom fell out of the nuclear industry.

The accident at the island did not precipitate the financial crises in the nuclear industry; it was only the final straw in a decade of growing political pressure that by 1979 had already darkened the financial future of nuclear power. And though much criticism of the NRC followed the Three Mile Island accident, the economic factors that jeopardized the future of the nuclear industry did not arise solely from governmental policy at the NRC or elsewhere. The continued and growing presence of political and ethical issues surrounding nuclear power also contributed to the present torpor evident in the Midland case. These issues are the domain of the public and of the political system as a whole, not of regulatory agencies.

ETHICS AND POLITICS IN NUCLEAR POLICY

When President Eisenhower in his "Atoms for Peace" speech set nuclear power on its developmental course, the major questions surrounding the technology were technical. Issues of cost, feasibility, and reactor safety had the highest priority, whereas the question of the desirability of developing and exploiting the new technology was seemingly answered. Thus, following the traditional path of policy making, the task of overseeing the project passed from the legislature and the president to administrative agencies—in this case the Department of Defense and the AEC.

This natural transfer in the policy process from political institutional settings to administrative ones parallels the progress in decision making from questions of value to questions of technical expertise. In other words, when confronting policy decisions in new areas like nuclear technology, the Congress and the president as representative offices must ask whether as a society we want, or should want, to proceed. Furthermore, they must ask what the consequences of going forward are likely to be and whether they are acceptable in light of other societal goals. These questions engage the preferences, values, and goals of a society. Only when they are addressed and resolved can the administrative question of "how" be asked.

Critics of our technological age sometimes claim that the worship of technology reverses this natural order in decision making. In the mad pursuit of development, they argue, we tend to ask first "do we know how" or "is it possible?" And if the answer is yes, the value question of "should we" rarely gets asked at all. In short, we make decisions, critics allege, according to the premise that "can implies should."[16] This complaint is pervasive today and has had much to do with the present state of nuclear power and policy. Whether the questions of value were ever asked before proceeding with nuclear development in the 1950s is by now a moot point. The significant fact in the 1980s is that the controversy over nuclear power is largely focused on these value questions, and the lack of consensus over their answers has stymied the continued deployment of the technology.

Siting controversies similar to the one in Midland involve essentially three ethical or value questions: The first centers on issues of social justice or equity; the second on demands for participation in the decision-making process; the third on attitudes toward risk. Because the question of risk in nuclear technology includes the dilemma of disposal of radioactive wastes, it will be discussed in Chapter 5. Each of the other two questions has many facets, and each has moved the nuclear debate in a different direction and with different consequences in several siting disputes.

Nuclear Justice

In the Midland case, the announcement of a site for nuclear plant construction began a series of events that eventually caused the project to be canceled because of public opposition. Opposition is common in siting decisions regarding nuclear plants, and disputes frequently accompany other public works projects like dams, airports, or, as we will see in Chapter 5, hazardous waste disposal sites. In fact, siting controversies pervade the private sector as well, as illustrated by public opposition to new shopping malls, apartment complexes, or low-income housing projects. The pervasiveness of these disputes perhaps witnesses to the basically conservative resistance of the individual to change, but more important to our discussion, it says something about the U.S. emphasis upon individual rights and about the perceived injustice of sacrificing them in the name of some amorphous public good.

Individuals' Rights. As stated in the Constitution, the U.S. political system is founded on the idea that all individuals equally have rights that are both "natural" and "inalienable." These rights are natural in the sense that people have them not because they have earned them or because the U.S. government grants them, but simply because all human beings have them. They are inalienable because no one can take them away from individuals without their consent. Furthermore, the main purpose of government is to protect and respect these rights, for it can not grant them nor can it take them away without the consent of the individual.

This dual conception of rights as natural and inalienable is basic to U.S. political life and is deeply ingrained in the minds of citizens. When citizens feel that their rights are not being considered equally with those of others or that their rights are being violated in the name of the good of other people (even all others in society), they claim that injustice is being perpetrated against them by society or government. Siting disputes grow out of this visceral understanding of the rights of individuals and the purpose of government. Stated simply, citizens opposed to construction of nuclear plants claim that their rights are not being given equal weight with the rights of others. Everyone else in society may indeed benefit from the siting of the plant (dam, airport), but those living closest to it must pay a higher price in terms of noise, risk of irradiation, or fear. Thus their rights are being violated without their consent, and this constitutes injustice.

The nature of the conflict in these instances is stated in various ways both by participants and by those who study them.[17] Legally, such disputes are questions of equity or fairness: Are some paying an unacceptably higher price than others for the same benefit? Constitutionally, the question is whether all are receiving equal protection as guaranteed in the fifth and fourteenth amendments. Philosophically, the question concerns the nature of justice as conceived on either utilitarian or absolutist (natural rights) grounds.

Utilitarianism. The advent of nuclear power has raised to public consciousness two questions of justice, both of which are philosophical but which are nevertheless apparent to all people affected by siting disputes. The first deals with whether we as a society define justice according to the principle of utility; the second focuses on the rights of future generations. The principle of utilitarianism states that just acts are those that maximize happiness for the greatest number of people, even if a minority of people experience some unhappiness as a result. The idea of natural rights is to some extent unalterably contrary to the principle of utility. Natural rights doctrine insists that some things simply may not be done to people, regardless of the benefit to someone else. This principle stands behind the U.S. legal guarantees of due process of law and protection of the rights of the accused. If indeed the construction of nuclear power plants violates the natural rights of some people, then even though utility results for everyone else in society, building the plant would constitute injustice.

Governments, particularly those that incorporate the idea of majoritarianism, necessarily make decisions on the grounds of utilitarian justice. Holding too strongly to the idea of natural rights or moral absolutism simply seems impractical in the real world because it implies that unanimous consent must accompany every decision. Yet people do have rights, and the Constitution says that the function of the U.S. government is to protect those rights from being violated. When some people claim that the majority is ignoring their rights in making its utilitarian decisions, as in the case of nuclear siting disputes, the question of justice is extremely difficult to resolve.

Justice Across Generations. A second issue involving justice raised by nuclear power is even more perplexing than that of utilitarianism versus natural rights. This is the question of justice across generations: What, if anything, do the living owe the not yet born? This issue is also raised in disputes over abortion and reproductive technology (see Chapter 8). In the context of nuclear power, the issue is what risks (in the form of radioactive waste) does the living generation have a right to impose on future generations. Complicating this question is the consideration that if nuclear power is not utilized, do the living have the right to exhaust supplies of fossil fuels that the generations of the foreseeable future would need to supply their energy needs?

Questions of justice and the balancing of rights seem to be—and frequently are—paralyzing to policy development. Decision making is difficult in an ethical environment; perhaps it was meant to be by the framers of the U.S. Constitution, who were wary of excessive governmental power. But when ethical questions are raised in the formulation or execution of policy, the courts naturally play an increasingly larger role in the policy process. This growing role is evident in nuclear policy, as the Midland case illustrates. To put it simply, when people feel their rights are being jeopardized, or they are looking for justice, sooner or later they go to court. And when they feel that way because of executive or legislative policy, the courts themselves become increasingly involved in setting policy.

Policy Making in Court. Policy making by the courts is intrinsically controversial within the U.S. political system, but when issues are redefined as ethical instead of merely technical, action by the courts has historically been equally inevitable. Judicial policy making is by no means a new phenomenon, and given the U.S. system of federal checks and balances and separation of powers, it is at least arguably a necessary one. In the area of nuclear policy the impact of judicial decisions has been pronounced and somewhat paradoxical. Though in retrospect the nuclear industry can be said to have never lost a major judicial battle, the effect of those battles has led to the present uncertain future of nuclear power.

Some of the judicial decisions regarding nuclear power have already been discussed. The Federal Appeals Court decision in the Midland case and the *Calvert Cliffs* decision imposed new regulations on the siting of nuclear power plants and altered nuclear policy in ways that have hindered the growth of the technology. But in other decisions the court has gone even further in its attempt to untangle the rights of all citizens affected by the siting of nuclear plants.

In the 1965 case of *Scenic Hudson Preservation Conference* v. *The Federal Power Commission,* the Supreme Court expanded its definition of *standing* to include environmental groups. The term *standing* refers to the designation of who can bring cases to court for a decision. Traditionally, only those parties directly affected by the actions of others are granted standing, but the court decided that the concern of environmental special interest groups over the impacts of nuclear plant siting constituted a right that could be claimed in court. Thus, the court expanded the number of groups that could take their case to court on the grounds that their rights had been violated by the construction of nuclear plants.

A second case went further in increasing the number of such individuals and groups. In *Environmental Defense Fund* v. *Hardin* (1970), the Supreme Court redefined the term *injury* for which reparations could be sought in court. Traditionally the judicial concept of injury referred to economic or physical damage suffered by someone who could seek compensation to be assessed in the courtroom. In this case, however, the court said injury could also include aesthetic or moral damage to the environment, to individuals, or to society as a whole as a result of disregard by the nuclear industry in its selection of sites and in its decision-making procedures.

These court decisions have affected the politics of nuclear power in two ways: (1) They have reinforced the ethical component of nuclear policy decisions, a component that has often led to the alteration or invalidation of nuclear policy; and (2) they have changed the politics of nuclear policy in an even more basic way by relocating it. No longer is policy made solely in the legislative or executive agencies at the state and local levels, or even in the courts. Rather, nuclear politics has moved to an arena populated by an enormous number of organized and unorganized groups of citizens, all claiming the right to participate in the policy-making process.

Nuclear Politics

On any issue, when citizens believe that their interests or welfare is directly affected, they are more likely to participate in the decision process. This participation is particularly evident in siting decisions, whether for dams, airports, shopping malls, or nuclear power plants. As we have seen, one reason for the public demand for participation is that these decisions raise questions of value choice, and U.S. political culture does not acknowledge "value" experts. All siting disputes raise issues of justice or equity, for all require that the perceived interests of those involved be acknowledged and balanced in a manner considered equitable or just. Siting disputes over nuclear plants have raised special balancing problems, however, because the type of interests involved are not usually considered in the cost/benefit equations of technological development.

In a capitalistic society like that in the United States, economic and technological development occurs in several contexts, both public and private. The history of nuclear power has shown a distinct federal prejudice in favor of private development by profit-making corporations serving the public interest through utility companies. Government has sustained this bias for private enterprise through actions like the Price-Anderson Act. Thus, nuclear development has proceeded on the basis of a profit equation indicating to investors and utility companies that money can be made through the exploitation of nuclear technology. In other words, the cost/benefit analyses of nuclear power have consistently recommended the relative cheapness of nuclear power as compared to power from other fuels and have therefore encouraged development.

Externalities. But the promises of cheap energy and profits have not been kept in the case of nuclear power. Part of this failure is the result of the increasing costs of litigation. However, the primary reason that nuclear power no longer seems economical is that the initial cost/benefit analyses failed to take into account certain costs now recognizable to the public, and this recognition has in turn prompted the increase in litigation. Although economists and social scientists have variously named these costs "disamenities," "diseconomies," or "spillovers," they are usually referred to as "externalities."

Externalities are costs that are borne not necessarily by investors or producers but by all members of society. They are somewhat hidden costs because they frequently affect negatively things that are not the property of any particular person or persons. For instance, an externality of burning coal to produce energy is pollution of the air. Since the air belongs to no one in particular, traditional cost/benefit analyses have not included in the cost equation of coal-generated energy the costs of cleaning up the air or the health costs of living with pollution. Since neither the utility nor anyone else can be said to own the air, the utility is under no obligation to include the costs of protecting air quality in its production of energy. If it did attempt to compensate society as a whole for air pollution, the utility's profit margin would likely disappear.

Externalities can be positive also, as are the benefits of living next to a beautiful and quiet public park, but in any case they are social effects not represented or articulated in the projections of profit-making enterprises. K. S. Shrader-Frechette pointed out some of the dilemmas of externalities.

"Externalities" are social costs. Because the social costs of producing certain goods, e.g., energy, are often not calculated, many persons wrongly assume that they do not exist. . . . This premise is widely accepted because "there are costs or benefits which accrue to society, but which are not included in the original contract between parties to an exchange. . . . Like negative externalities, positive externalities are also involuntarily imposed.[18]

The presence of externalities accompanying technological development is partially responsible for the tremendous growth in governmental expenditures and regulation since the start of the twentieth century. If externalities are costs borne by society as a whole, then government—which represents society as a whole—would legally be concerned with them. However, the governmental (and therefore political) concern with externalities arising from private endeavors means that those endeavors become themselves political, as all members in society recognize the stake they have in them and seek to have their interests recognized. This politicization occurs in siting disputes.

Advocacy Politics. By their very nature externalities encompass a different kind of interest than that usually represented by "special interest groups" in siting disputes or in U.S. politics generally. Externalities are the interests of everyone in society. Thus, those who claim to be concerned with externalities consider themselves to be protecting not their own but the public interest. In the case of nuclear power, many so-called public interest groups have been active in the debate over siting, as evidenced in Midland. Some of these groups are local, like the Saginaw Valley Nuclear Study Group; others are national, like the Public Interest Research Groups (PIRGs) and the organizations begun by consumer activists like Ralph Nader. These groups participate in what Constance Cook called "legal advocacy" and "advocacy politics": politics characterized by legal disputes in civil court and by lobbying at every level of government, all in the name of protecting the public interest from the burden of externalities.

Advocacy politics permeates U.S. politics today, particularly in areas of science and technology policy. In one sense advocacy politics is not new, since public interest groups are represented by lobbies in the style of Dow Chemical, General Motors, or any other private concern. In another sense, however, public interest groups are special: They seek to persuade as many citizens as possible that their interests are being represented by the public interest lobby. Public interest groups seek to convince all citizens that they have a stake in the externalities of nuclear power and that all the interests of any one citizen are affected by the externalities of nuclear power. Because of this double involvement in the external costs of nuclear power, public interest groups urge all citizens to become involved in the debates over siting and regulation of nuclear energy.

The success of advocacy politics can be seen in the increase of citizen and public interest group involvement in nuclear policy. A negative aspect of this success has been the decline of the nuclear industry. But is there a positive side to the increased involvement of citizens in the development of policy and of the policy environment that together have so decimated the industry? Although it is often argued that increased participation by citizens in a democracy is an indisputable asset, not everyone would agree.

Certainly in places like Midland advocacy politics has damaged the nuclear industry and therefore threatened long-term energy policy. As in many other siting disputes, those over nuclear siting have led to confusion over policy and in some cases have overturned policies that seem admirable and sensitive to an array of issues and interests. Since the primary externality of nuclear power is additional risk to the public, we need to examine the public attitudes toward risk before we can assess the rationality of policies made in the name of the public interest but having the additional effect of threatening our energy future.

NOTES

1. From a report on hearings before the California State Committee on Resources, Land Use, and Energy carried in the *New York Times,* June 2, 1976, p. 1.

2. C. Wright Mills, *The Power Elite* (New York: Oxford University Press, 1959), p. 5.

3. Constance Ewing Cook, *Nuclear Power and Legal Advocacy* (Lexington, Mass.: D. C. Heath, 1980), p. 49. Much of the description of the following Midland case is derived from Cook's fine work.

4. Cook, *op. cit.,* p. 65.

5. Quoted in Cook, *ibid.,* p. 83.

6. Data derived from *Time* 123, no. 7 (February 13, 1984), pp. 34–45.

7. *Ibid.*

8. Robert C. Williams and Philip L. Cantelon, eds., *The American Atom* (Philadelphia: University of Pennsylvania Press, 1984), p. 71.

9. See K. S. Shrader-Frechette, *Nuclear Power and Public Policy* (Boston: D. Reidel, 1980), pp. 8–10, for a discussion of this aspect of U.S. civilian nuclear power.

10. *Ibid.*

11. Quoted in Williams and Cantelon, *op. cit.,* pp. 296–301.

12. See Donald R. Whitnah, *Government Agencies* (Westport, Conn.: Greenwood Press, 1983), pp. 395–399, for a discussion of the Atomic Energy Commission's demonstration project in the 1950s.

13. *Ibid.,* p. 396.

14. Cited in Shrader-Frechette, *op. cit.,* p. 11.

15. For an excellent account of the events at Three Mile Island, see Phillip L. Cantelon and Robert C. Williams, *Crisis Contained: The Department of Energy at Three Mile Island* (Carbondale, Ill.: Southern Illinois University Press, 1982).

16. This criticism is voiced in a vast amount of literature that seeks to cast light on the problems of modernity. Perhaps the most famous study is that by Jacques Ellul, *The Technological Society* (New York: Knopf, 1964).

17. See, for instance, Shrader-Frechette, *op. cit.*; Robert E. Goodin, *Political Theory and Public Policy* (Chicago: University of Chicago Press, 1982); Cook, *op. cit.*; Steven Ebbin and Raphael Kasper, *Citizen Groups and the Nuclear Power Controversy* (Cambridge, Mass.: MIT Press, 1974); and Steven L. Del Sesto, *Science, Politics, and Controversy* (Boulder, Colo.: Westview, 1979).

18. Shrader-Frechette, *op. cit.,* pp. 108–109, passim.

SELECTED READINGS

Berger, J. J. *Nuclear Power.* New York: Dell, 1977.

Cook, Constance Ewing. *Nuclear Power and Legal Advocacy.* Lexington, Mass.: D. C. Heath, 1980.

Del Sesto, Steven L. *Science, Politics, and Controversy: Civilian Nuclear Power in the United States, 1946–1974.* Boulder, Colo.: Westview, 1979.

Ebbin, Steven, and Raphael Kasper. *Citizen Groups and the Nuclear Power Controversy.* Cambridge, Mass.: MIT Press, 1974.

Miller, Saunders. *The Economics of Nuclear and Coal Power.* New York: Praeger, 1976.

Novick, Sheldon. *The Electric War.* San Francisco: Sierra, 1976.

Shrader-Frechette, K. S. *Nuclear Power and Public Policy.* Dordrecht, Holland: D. Reidel, 1980.

Williams, Robert C., and Philip L. Cantelon, eds. *The American Atom.* Philadelphia: University of Pennsylvania Press, 1984.

Willrich, M. *Global Politics of Nuclear Energy.* New York: Praeger, 1971.

5

HAZARDOUS WASTE: ASSESSING RISK

By its nature, social life requires that individuals live with the consequences of other people's actions. These actions may benefit each individual, but they also impose certain costs that individuals must bear for the preservation of the common social life. In Chapter 4, we spoke of these costs of collective life as externalities and noted that all technological growth imposes new external costs that society must recognize and include in its policy structure to preserve the safety and health of citizens. Life in modern technological societies is characterized by many externalities. In this chapter we will look at two of the most prominent social costs born by every citizen: the high price of radioactive waste disposal and the risk that accompanies it.

This chapter focuses primarily on the concept of risk as a byproduct and external cost of scientific and technological advance. The idea of risk is difficult to grapple with, however, for it seems to have two distinct manifestations: its objective character (how much risk actually exists) and its subjective character (how much risk people believe exists). In terms of policy making aimed at controlling the degree of risk in society, it is unclear which of these two characters should be the primary focus. Because of the abstract nature of the concept, we will illustrate the idea of risk by describing a tangible policy area in which risk plays a prominent part: waste disposal policy.

U.S. society has been characterized by some as the "throw-away society," and statistics indicate that Americans produce a staggering mass of waste. The category of waste covers many types of refuse, from aluminum cans and waste paper to hazardous petrochemicals and radioactive materials. Much of the writing about the problem of waste disposal in the United States has focused on the problems of the so-called hazardous waste of chemical and petroleum waste products.[1] We will concentrate on radioactive waste disposal as a policy area, not because it is necessarily a more pressing problem than that of chemical waste but because it has received less attention both from scholars and policy makers and because it best illustrates the problematic and confusing nature of risk.

RADWASTE IN NEW MEXICO

In the southeast corner of New Mexico, midway between the small towns of Carlsbad and Hobbs, lie 27 square miles of desolate, sagebrush-covered property chosen in 1980 to begin living up to its description as "wasteland." In that year the federal Department of Energy (DOE) selected the tract as the first site for its radioactive waste (radwaste) disposal project begun in 1976 and known as the Waste Isolation Pilot Project (WIPP). The area around Carlsbad was chosen by WIPP directors because it seemed to provide both the geological environment and the political environment thought to be essential for successful development of a radwaste installation.

The terrain around Carlsbad is barren and largely uninhabited. These aspects appealed immediately to WIPP scientists because they assumed that a radwaste repository should be located as far as possible from a population center. However, the isolation of the site was a secondary reason for selection. The primary reason was that the subsurface geology of the area was exactly that which, since the 1960s, scientists regarded as ideal for the containment of radiation given off by radioactive waste products. Deep below the sand and sagebrush were huge beds of salt with few drill holes from past mining. Scientists consider salt beds the safest burial medium for radwaste because such beds indicate the absence of underground water nearby in aquifers or springs so that radiation from the radwaste could be contained without contaminating surrounding areas. Furthermore, salt beds withstand intense heat well and are relatively "plastic," meaning that any cracks caused by the intense heat given off by the decomposing radioactive materials would quickly seal themselves. For geological reasons then, the site around Carlsbad seemed ideal for the model WIPP site.

Similarly, the political environment in New Mexico seemed well-suited for construction of a major radwaste facility. New Mexicans are accustomed to living in close proximity to nuclear energy installations. Although nuclear reactors generate electricity in many other states, New Mexicans point with pride to their Los Alamos installation—where the first atomic bomb was created during World

War II—and to other weapons factories around the state. In short, the directors of WIPP felt that because New Mexicans were comfortable around radioactivity they would not cause undue political problems over the construction of the site.

In 1981, however, the geological and political appeal of the New Mexico site began to lose some of its luster. Late in November exploratory drilling revealed a pocket of salt brine caused by the mixture of salt and water. This feature was precisely what scientists in the project hoped would not be found: Water in the underground salt meant that the containment of radioactivity could not be guaranteed. This discovery did not mean that the site would be immediately abandoned, but it did heighten the political controversy begun earlier in the year.[2]

Beginning in 1980 the politics surrounding the site selection became confrontational, when President Jimmy Carter altered the mandate of the site. The original WIPP purpose was to find a permanent storage site for low-level and transuranic waste products from defense programs. (These designations of radioactive wastes will be detailed in the next section; such wastes do not pose severe environmental or safety hazards when compared to many forms of commercial radioactive wastes.) Because of growing radwaste disposal concerns, President Carter ordered WIPP in 1980 to plan on using the site as the dumping ground for approximately one hundred commercial-reactor spent-fuel assemblies and other extremely high-level radioactive waste products. Because of the absence of repositories for these wastes, New Mexico was alarmed at the prospect of becoming the prime dumping ground for the nation's high-level and extremely hazardous radioactive refuse.

New Mexico was not the only injured party in the Carter decision. Allowing the Carlsbad site to accept commercial and high-level waste would bring it under the purview of the Nuclear Regulatory Commission (NRC) which has statutory authority in licensing radwaste projects. Not surprisingly, the DOE was adamantly opposed to forcing the Carlsbad project to accept commercial waste because this decision would transfer full authority for the project from the DOE to the NRC. Furthermore, the Defense Department opposed the NRC licensing because DOD's waste disposal procedures had been exempt from NRC regulation since the 1950s. After a complicated turf battle between Congress, the president, the DOE, and the NRC, President Carter canceled the project in February 1980. However, Congress restored funds for the project later in that year, and New Mexico still awaited assurances that it would not become the repository of the nation's unwanted waste.

On May 14, 1981, New Mexico sued the DOE, claiming that the department had "refused to agree to a legally enforceable document to resolve" remaining disputes over the type of waste that the Carlsbad project would be forced to accept.[3] According to a federal law passed in 1979, the DOE was to consult and cooperate with states in which disposal sites had been proposed before any construction would begin. Consultation between the DOE and New Mexico had dragged on for a year without assurances that high-level waste would not be stored at the WIPP site. Less than two months after the suit was filed, the state received these assurances and reached an agreement with the DOE to continue the construction at the site; drilling of the twenty-one one-hundred-foot shafts began.

With the discovery of the brine pockets all construction was halted. After much bureaucratic political wrangling and ill will, federal legislation and public controversy, the presence of water in the salt beds seemed to doom the project once again. In January 1983, the DOE decided to move the project to a new site (as yet undetermined), and the Carlsbad site was abandoned. Though President Reagan had made WIPP a priority of his first administration and had added funding to it on the third day of his presidency, the program had made no new site selection by 1986. The politics and geological requirements of radioactive waste disposal had doomed another potential repository site, and no new alternatives for disposing of the nation's mounting stock of radwaste have yet appeared.

RADWASTE IN THE UNITED STATES

The failure of WIPP highlights the dangers and dilemma of radioactive waste disposal in the United States. For forty years the country has pursued the development of nuclear energy without determining how to dispose of its hazardous byproducts. Alvin Weinberg, past director of Oak Ridge National Laboratory, described the problem of radioactive waste as a situation in which "nuclear scientists have made a Faustian bargain with society." Acknowledging that he and his colleagues offered "energy that is cheaper than energy fossil fuel, and when properly handled is almost nonpolluting," he concluded that "the price that we demand of society for this magical energy source is both a vigilance and longevity of our social institutions that we are quite unaccustomed to. . . . This admittedly is a significant commitment that we ask of society."[4]

As the Carlsbad case shows, even the most politically and scientifically popular proposed solution to radwaste disposal—deep disposal in geological salt beds—still carries unsolved technical and political problems. In spite of federal legislation and continuing site exploration, it is estimated that no new disposal installations will be ready for operation until the mid-1990s. Meanwhile, the mountain of radioactive wastes of all varieties continues to grow, and the prospects for removing it require faith in both U.S. technical prowess and political institutions that forty years of failure does not encourage.

Disposing of radioactive materials safely poses several perplexing issues of risk management. The risks derive from the nature of the material itself and the length of time over which vigilance by political and regulatory institutions is demanded. Before we examine the political and policy issues raised by radwaste disposal, we must look at the nature of the material itself to understand the unique risks it poses to public health both now and in the future.

Types of Radioactive Waste

Radioactive wastes range from gloves tainted with radioactive elements to the spent-fuel assemblies of nuclear reactors. All these products share a potential health hazard that renders them, in the terms of the NRC's official definition, "materials which are of sufficient potential radiological hazard that they require special care and which are of no present economic value to the nuclear industry."[5] Radioactivity is itself a mystery to many. Put simply, radioactive materials are those tainted by a radioactive variant (or isotope) of an element. Radioactive isotopes are unstable and to stabilize themselves emit high-energy particles as well as heat. Together these emissions are known as radiation. As atoms of radioactive isotopes go through this process, they are referred to as "decaying." Radiation is harmful to organic tissue like the human body, and in sufficient quantities it causes mutation of cells subjected to particle bombardment, resulting in cancer or, in extreme cases of exposure, rapid death.

Classifying the many varieties of radioactive waste is difficult. Current classification schemes adopted by the NRC are based on where in the nuclear generation cycle the waste is generated rather than on the degree of radioactivity or the time required for the waste material to decay into inert elements (measured in half-lives). This type of classification scheme is itself controversial, since, as one government report stated, classification by location in the nuclear cycle "does not necessarily provide a key to potential radiation hazards."[6] In any case, six different types of radwaste are recognized by the NRC, and they are listed here in order of their production in the nuclear cycle: uranium mill tailings, low-level wastes, transuranic wastes, spent-fuel wastes, high-level wastes, and decommissioning wastes. Each of these poses special radiation hazards and policy difficulties.

Before uranium can be used as fuel for nuclear reactors, it must be refined at a mill to remove trace elements and purify it. The products of this process are fuel-grade uranium and solidified waste elements that have a sandlike consistency (known as mill tailings). As radioactive waste they emit the smallest amount of radiation of all radwaste products, but they are nevertheless hazardous and have a volume of almost twenty times that of all other radwaste types combined.[7] In 1980 the Oak Ridge inventory counted 2.955 billion cubic feet of radioactive tailings at a total of forty-five active and inactive uranium mills, an amount that would cover 106 square miles at a depth of 1 foot. The volume is growing each year at the rate of 141 cubic feet.[8] In the past decade, thousands of tons of the tailings have been inadvertently combined in the concrete of building foundations, and the EPA estimates that at least 3 million tons of tailings have blown into the Colorado River alone, where they enter the food chain and are concentrated at each successive step in the chain. At present little disposal space is available for even a small percentage of the volume of tailings requiring disposal, and the simple transport of tailings has proved difficult.

Low-Level Waste

Low-level radioactive wastes constitute the largest variety of waste products. Originally, NRC defined low-level radwaste as all radioactive materials not produced during the reprocessing of nuclear fuel. According to the Oak Ridge inventory, 85 million cubic feet of low-level wastes are buried at federal and commercial burial sites, an amount that would cover more than a square mile at a depth of 3 feet. In 1979 the United States produced 2.9 million cubic feet of low-level wastes, and this figure is estimated to rise to over 5 million cubic feet per year by the end of the 1980s.[9] Also, since 1947 an unmeasured but sizable percentage of all low-level waste has been put in barrels and dumped by the U.S. military and by other nations into the Atlantic and Pacific Oceans. Evidence gathered by the NRC and other agencies indicates that much of the radwaste is now leaking out of its original containers and polluting the ocean floors.

Low-level wastes are not necessarily low-radiation wastes. Some low-level waste facilities have accepted large quantities of strontium 90, cobalt 60, cesium 137, and plutonium, all of which are extremely radioactive and have half-lives of thousands of years.[10] By 1986 only two low-level waste sites were still operating in the United States:

one at Hanford, Washington, and the other at Barnwell, South Carolina. Both are scheduled to close within four years.

Transuranic Waste

In 1970, the Atomic Energy Commission designated a new category of radioactive waste called transuranic waste. Technically, transuranic wastes are those materials containing elements with atomic numbers higher than uranium. All these radioactive elements are human created; they are the products of defense-oriented nuclear projects or products potentially useful for defense purposes. The most well-known example is plutonium. Thus, the designation of these products in 1970 was meant to encourage their recycling and their disposal by methods that would allow future reclamation.

Because of the possible recycling of transuranic wastes, since the 1950s these products have usually been stored with low-level wastes in shallow ground trenches or in barrels and crates buried beneath 3 feet of earth. Some transuranic wastes like plutonium and strontium are extremely volatile and have half-lives of thousands of years. Their volatility has in fact led to near-disasters in twelve shallow waste sites since 1959. At these sites events occurred that are called "inadvertent criticalities" or "criticality excursions," which is bureaucratese for nuclear chain reactions—in other words, nuclear explosions. Simply put, plutonium wastes buried in shallow trenches reached critical mass and started chain reactions characterized by explosions and the release of varying amounts of radiation.[11] These events, which occurred at waste sites operated by the government for the Defense Department, have never been widely publicized in the United States. No injuries have ever been reported from the criticality excursions, though there is wide speculation that such an excursion in a transuranic waste site in the Soviet Union in the 1960s may have destroyed 10,000 acres of land in Siberia along with all vegetation and animal life.

Estimating the amount of transuranic waste in the United States is difficult because it is a defense-related material. Nevertheless, at least 26 million cubic feet of transuranic waste were certainly put into retrievable storage before 1970 in eight government and five commercial sites.[12] Furthermore, the Environmental Protection Agency estimated in 1978 that the volume of transuranic wastes was growing at an approximate rate of 225,000 cubic feet per year.[13]

Spent-Fuel Assemblies

Perhaps the most dangerous form of radioactive waste, as well as the most perplexing to deal with, has never been officially designated by Congress as waste. This form consists of the spent-fuel assemblies of nuclear reactors, both civilian and military. These assemblies are made of rods of uranium pellets that act as fuel for the reactors, along with rods made of particle-absorbing metals that control the speed of the nuclear reaction and are themselves made radioactive in the nuclear process. Fuel assemblies emit radiation in levels that would be lethal to humans in ten minutes at a distance of 10 yards.[14] Fortunately, this radiation can be shielded by water; thus spent-fuel assemblies are immersed in huge pools on site at most nuclear plant installations. The waste disposal problem for spent-fuel assemblies is the growing lack of space for immersing spent-fuel assemblies.

By 1981, almost thirty thousand spent-fuel assemblies were being stored on site in pools at nuclear power plants in the United States. By 1990, twenty-eight of these plants will no longer have room to store the assemblies. By 1986, the federal government had not yet contracted for a permanent storage site for the assemblies. When and if a site is chosen, its construction will take years to complete—years after the 1986 crisis date set by the Department of Energy.[15]

The absence of a permanent solution for the disposal of spent-fuel assemblies also increases the amount of high-level radioactive wastes because most high-level wastes, by the NRC designation, are liquids used to cool nuclear reactions. Cooling water is used throughout the reactor operating process and also in pools holding spent assemblies. Except for spent-fuel assemblies, which are not officially designated as radwaste, this water contains the highest levels of radiation of all radioactive wastes. Thus it is extremely hazardous. By 1982 commercial nuclear reactors were estimated to have produced more than 8,000 metric tons of high-level waste and military reactors even more than that amount, though the military total is secret.[16] At present, all high-level radioactive wastes from commercial reactors are being kept in liquid containers or in capsules at the two remaining high-level waste treatment centers in Hanford and Barnwell.

D&D Waste

The final form of radioactive waste for which methods of disposal must be found is the nuclear reactors themselves. The life of a nuclear reactor does not exceed forty years, after which time the materials used in its construction are so radioactive and brittle from radiation bombardment that they must be disposed of as waste. This type of waste is known as decommissioning and decontamination (D&D) waste. These reactors include those used in commercial electricity-generating plants, nuclear submarines, atomic particle accelerators, uranium mills, and nuclear fuel fabrication plants. The total number of reactors either now classified as waste or soon to be so is almost fourteen-hundred units.

Nuclear reactors are sizable pieces of machinery. Disposing of them in a way that limits the amount of radiation given off during their thousands of years of decay requires large masses of shielding material, but the type of material is not yet agreed upon. Though the U.S. Navy has requested to be allowed simply to scuttle its nuclear submarines in the ocean when they are decommissioned, the resulting effects on aquatic life are difficult to estimate. Although some reactors have been chopped up and shipped to high-level waste storage centers, again space problems prevent this from being a permanent solution. The currently favored plan is simply to lock up the reactor buildings of commercial reactors and post warning signs and guards for the foreseeable future—a future that consists of thousands of years. One other solution is to cover decommissioned reactors with huge concrete mausoleums, entombing them in several feet of cement that, experts hope, is impenetrable by radiation.

RADIOACTIVE WASTE POLICY SINCE 1977

In 1973, Senator Howard Baker of Tennessee declared that "the containment and storage of radioactive wastes is the greatest single responsibility ever consciously undertaken by man."[17] We can recognize the extent of that responsibility now that we have reviewed the different varieties of radwaste and the special problems accompanying each. Although the dilemmas of radwaste disposal are impressive, they have a short history—roughly forty years since the creation of the first atomic device in the Manhattan Project. This brief period for recognizing and dealing with the dangers of radioactive waste is considerably longer than the policy history of radwaste. As a focus for policy making, the issue of disposal of these wastes was not even recognized till the early 1970s.

Throughout the 1970s scientists expressed their concern in scholarly papers and in congressional testimony about the need for radwaste disposal techniques, but not till 1977 did the federal government take any action.[18] In April of that year, President Carter created the Interagency Review Group (IRG) and charged it with developing a method of radwaste disposal. In 1978 the group reported to the president that if construction began immediately, the first model for a high-level disposal facility could begin operation by the mid-1990s.[19] Of course, construction did not begin immediately because it was contingent upon congressional legislation regarding plans for a facility and for its site.

Congress began action on the IRG report in 1980 and quickly realized the beguiling problems of disposing of radioactive waste. First, the time interval required for decay of some radwaste is almost incomprehensibly large: Some high-level wastes have half-lives of thousands of years. Because U.S. political institutions are only a little more than two-hundred years old, the idea of creating policies that must remain in effect essentially for eternity is a mind-boggling proposition. Of course, storage of high-level wastes under any long-term policy immediately raises questions of justice across generations, as noted in Chapter 4. As Harvey Kendall of the Massachusetts Institute of Technology (MIT) reminded the Atomic Energy Commission in 1973, high-level waste disposal gives rise to "a new kind of risk-benefit calculation where we get the benefit now and hand the risks on to other generations, which will get no benefit at all."[20] Unless these considerations are taken seriously, making policy for radioactive waste disposal is shortsighted at the very least; at most it seems to many to exhibit what the ancient Greeks called *hubris*— a false, often fatal pride in the ability to control one's life and environment. The AEC acknowledged the dangers of this character flaw when it noted that "fission technology requires that man issue guarantees on events far into the future, and it is not clear in most cases how this can be done. Institutional arrangements do not exist and never have existed to guarantee the monitoring of or attendance upon storage facilities over a millennium."[21]

The second problem of radwaste disposal is the risk of leakage and contamination that disposal poses for the environment. This fear is manifested in the profound reticence of states and localities to accept a disposal site within their borders. From a practical standpoint, this second problem has done more to stall policy solutions for radwaste disposal than the first. Beginning in 1980, Congress began to see the effects of this second problem in its relations with state and local governments, as evinced in the WIPP case in New Mexico.

In December 1980, Congress passed its first piece of legislation regarding radwaste disposal, in which it essentially tossed the disposal problem back to the states by issuing two ultimatums. First, the legislation authorized several states to deny access to existing disposal centers to operators from other states unless at least three additional disposal facilities were constructed by 1986. Second, the legislation directed states to solve their own disposal problems by establishing regional coalitions or compacts that would jointly construct their own disposal centers. The legislation also allowed states that entered into regional compacts to exclude nonmember states from using their disposal sites.[22]

In the years since 1980, the primary purpose of the legislation has not been met. Although some states have established regional compacts, no new disposal facilities have been built. However, a secondary effect of the legislation has been the increasing competition between states for the right to dispose of their radioactive waste at existing sites and a renewal of the Balkanization syndrome among states described earlier. The present manifestation of this syndrome is rather paradoxical: Because the number of radwaste facilities continues to decrease, no state can easily dispose of its dangerous refuse; yet, instead of cooperating with neighboring states, many states have entered disputes over the transport of radioactive wastes on roads and waterways within their borders. Ironically, when the final two radwaste disposal sites close by 1990, these transport disputes will become moot because radioactive wastes will then be on a road to nowhere.

Worsening relations among states and between states and the federal government over radwaste disposal led Congress in 1982 to pass additional legislation effecting construction of new disposal facilities. This legislation now constitutes the only federal policy regarding radwaste disposal. The National Nuclear Waste Policy Act (S. 1662) was finally approved by both houses of Congress on December 20, 1982, after much debate over the issue of state rights in siting radwaste facilities. The legislation mandates the Department of Energy to select two new sites for disposal facilities by 1989. The legislation includes a provision allowing a state to veto any site selected by the DOE within that state, a veto that would be upheld unless overridden by both houses of Congress.[23] This provision gives any state selected for a radwaste facility great authority to determine whether indeed it will allow construction. By 1985, no state selected by the DOE's initial screening of potential sites had agreed to permit construction.

The states' reluctance to house a radwaste depository can be explained by the known hazards of radioactive materials. Unfortunately, this rationale does not aid in making policy to deal with U.S. disposal needs. Instead, an understanding is needed about the base factor beneath this state reluctance—the public's attitudes toward risk in radioactive waste disposal. Those attitudes are frequently unrealistic because they are not based on accurate assessments of the amount of risk that actually exists. Nevertheless, they can effectively block policy on waste disposal, particularly because of the veto stipulation in the 1982 legislation. Thus, it is necessary to explore the entire area of risk and risk assessment as a necessary prelude to policy formation for radioactive waste disposal.

RISK AND RESPONSIBILITY IN POLICY MAKING

In the United States, risk taking is a factor of everyday life met with both fear and enjoyment. The free-enterprise system places a positive value on the willingness to take risks to achieve success, and the U.S. political theory of liberalism stresses the individual's ability and right to rationally cope with risk without undue interference by government. Yet all governments are exclusively charged with alleviating some kinds of risk in society, for example, those arising through crime or war. Thus the policy environment for dealing with risk is intrinsically filled with dilemma and paradox for those charged with making public policy. The long legislative and policy battles over Social Security, the 55-mile-per-hour speed limit, the drinking age, and automobile seat belts are examples of this curious paradox.

Modern technological society is undeniably safer than any other in history. The average life span of Americans continues to increase, and because of countless medical and technological innovations, the risk of contracting many diseases has lessened. Yet Americans are acutely aware of the risks of living in our time. Psychologists point to the unconscious fears of many young people concerning nuclear war, and the threats of crime and accelerating technological innovation bring new and uncomfortable feelings of living at risk.

Many of the new feelings of risk in modern society are simply the result of increased knowledge. The more we learn about ourselves and the world, the more risks we are able to identify. Knowing that risks exist, however, does not necessarily make them easier to eradicate. In fact, sometimes the more we know about environmental risks the more persistent they become as policy problems. Although awareness of the alphabet soup of toxic human-created chemicals in our environment (for example, EDB, PCB, PBC, DDT) has made citizens notice health risks, it has not made it any easier to formulate policy to protect the public health. Furthermore, the knowledge that we have made or chosen new risks frequently causes us to feel guilty—making policy formulation more difficult. The *New Yorker* cartoon of a woman at a lunch counter reading a menu consisting of entrees followed by their possible cancer-producing effects is a commentary on the public consciousness of self-assumed risk that often turns decisions about risk taking into issues of personal freedom and moral choice.

The same transformation affects large-scale technological policy as well. For example, on one hand, the risks of waste materials are known and largely undisputed, as is the need for policy in this area. On the other, the varying attitudes concerning the attendant risks make these policies difficult to formulate and implement because of claims of individual rights and free choice.

Attitudes Toward Risk

In the policy process concerning risk, the most important factor may not be the actual risk involved but the attitudes of citizens toward that risk. This emphasis, particularly true in democratic societies of course, points out the twofold nature of risk assessment as a step in the policy process. Actual (or objective) risk may be broadly defined as the measure of the probability of occurrence and severity of negative effects. Risk is always a measure of something, and it is always a stochastic or probabilistic measure, subject to some degree of uncertainty. Thus the actual measurement of risk is by definition somewhat imprecise and subject to interpretation.

The subjective side of risk measurement is even more imprecise because it involves the measurement of people's attitudes toward risk—a side of risk assessment and management obviously crucial in making policies that involve risk in a democratic society. As Nicholas Rescher pointed out, "perceived or subjective risk is something rather different from risk itself, [and] although the risk of various outcomes is a wholly objective, ontological issue of how things stand in the real world, the subjective side of risk is nevertheless unavoidable when addressing the issue of the appraisal and management of risks."[24] Basically, both aspects of risk present measurement difficulties because they involve the human element. Human beings create risks and hazards and the schemes for managing them, as well as hold widely divergent attitudes toward the risks and the policies generated.[25]

The objective reality of risk is further complicated by two characteristics that all risks manifest. First, risks are not absolute: They can be alleviated by preventative action. Consequently, measures of objective risk can be altered either upward or downward, depending upon sometimes unforeseen factors like the unpredictable responses to risk manifested by people. Second, risks compete with each other to present adverse effects. Although this sounds a bit strange and anthropomorphic, it actually is not. Risks compete with each other in the sense that if one risky event actually comes to pass, others by definition cannot. For example, the risk of death to an individual in either an automobile or nuclear accident is very real; yet if an individual has one accident he or she obviously cannot have the other—one can only die once. These aspects of objective risk also affect the subjective side of risk: If one lives with the risk of imminent

**Phillips Memorial
Library
Providence College**

death because of one feature in life (e.g., proximity to a missile silo in Kansas), fear of risks from other causes (e.g., smoking) may decrease.

Like objective risk, subjective risk is relative but to a greater number of factors. Attitudes toward risk are relative to perceived benefits and may change as personal values change or as familiarity with risk is extended. Generally, a host of factors affect subjective attitudes toward risk, which is why making a policy that distributes risk is so difficult. When confronted with a decision over the siting of a radioactive waste disposal facility, for example, policy makers must attempt to assess the subjective attitudes toward the risk that the facility presents. This attempt is warranted not only because democratic principles demand it, but because, as we have seen, not assessing risk attitudes can result in the failure of the policy.

Measuring Risk

Measuring subjective risk for policy making is usually done according to one of two models that the Congressional Research Office has identified. One uses either the Revealed (or Implied) Preference Model or the Expressed Preference Model.[26] The Revealed Preference Model begins by assuming that society has achieved an optimal and accepted balance between risks and advantages and then measures how much a new risk would affect that balance. One first ascertains the attitudes of citizens toward a specific amount of objective risk and then uses this as a basis for predicting their attitudes toward a specific amount of new objective risk.[27] The Revealed Preference Model is exemplified in the following policy statement by the Atomic Industrial Forum in 1976.

> The Nuclear Regulatory Commission has recognized an acceptable level of risk, at least for regulatory purposes, in granting permits and licenses. While this level of risk has not been specifically quantified, the Reactor Safety Study now provides a benchmark for comparison. With this background, new issues can be assessed by judging whether these issues impact significantly on the plant risk envelope as determined in the Reactor Safety Study. If an issue can be shown not to affect significantly this risk, then design alterations additional to the vintage plant design analyzed in the Reactor Safety Study could not be justified.[28]

Relying on experience as in the Revealed Preference Model has an intuitive appeal; yet several problems arise with this approach. First, the benefits of different risky activities in the past and present may not be directly comparable. Second, certain benefits do not readily

Phillips Memorial
Library
Providence College

admit of quantification at all. For instance, ascribing a numerical or monetary value to the benefit of clean air is difficult. Third, the Revealed Preference Model ignores the fact that the acceptance of risks is not necessarily transitive. The fact that the same number of people may die in an inadvertent criticality at a radwaste facility as die yearly on the nation's roads does not make it logical for citizens to accept the additional risk of a nearby radwaste facility even though they accept the risks of automobile travel. This final problem with the Revealed Preference Model has turned policy makers to the second model of subjective risk assessment.

The Expressed Preference Model relies on citizens' actual answers to questions about their willingness to accept new risk. This model uses questionnaires and modern polling techniques, and thus it is susceptible to the usual skepticism regarding such techniques, as well as to the criticism that people do not necessarily respond the same way on questionnaires as they would in actual risk situations. Nevertheless, by using this model many factors that affect the subjective attitudes toward risk have been identified. The Expressed Preference Model has indicated that people's attitudes toward additional risk are likely to be positive if the risky activities

- are voluntary as opposed to involuntary
- are avoidable
- are controllable
- are familiar
- are well understood
- are not dreaded
- present effects that are remote
- present chronic rather than catastrophic effects
- present immediate rather than delayed effects
- present effects that are not certain to be fatal

At the policy-making level, government agencies use these models to assess the degree of subjective risk new policies might present. For science and technology policy, this assessment is done primarily at the Environmental Protection Agency (EPA) and the Office of Technology Assessment (OTA). The OTA is a congressional agency, rather than an executive one, and thus it occupies a somewhat unique place in the federal bureaucracy. Established by Congress in 1972, the OTA was intended to provide "an early warning system for Congress, so that Congress can consider the social . . . impacts of technological advances . . . before these effects are upon us."[29] OTA has a mixed record as an advisory agency for Congress in the latter's attempts to make policy for new technologies and for the risks they present. The problems with policy making by the OTA and Congress are evident in the effects of the 1980 legislation concerning radioactive waste disposal site selection; they point to a basic failing in U.S. policy dealing with the distribution of risk as a part of science and technology policy.

Public Acceptance of Risk

As stated, U.S. political ideology and economic principles posit that risk taking by individual citizens carries a positive value in that accepting risk may bring certain rewards. Because of this deep-seated acceptance of risk as a part of political and economic life, Americans would seem to be uniquely willing to accept policies like the 1980 radwaste legislation that seek to distribute the risk of proximity to a radwaste disposal. Why then do such policies fail because of a lack of public acceptance based on the fear of risk? The answer lies in the fact that these policies ask citizens to accept risk without the promise of reward. In so doing, the policies demand a larger sacrifice by some citizens (who must accept a larger portion of risk) without offering adequate inducements to convince them that the additional risks are worth taking. Thus, the basic U.S. principle of risk acceptance is abused when legislators ignore its most persuasive element: the possibility of gain.

In our discussion of energy policy, we saw that President Carter's energy program failed because he was unable to convince Americans to accept sacrifice. Americans are schooled in the political theory of liberal individualism; thus, calls for sacrifice by some individuals must be accompanied by convincing reasons that making the sacrifice is worthwhile for themselves, not merely for society as a whole. The same argument is true for policies that seek to convince some citizens to accept the risk of living in close proximity to a radwaste disposal facility. Those citizens would not accept the sacrifice of living with additional risk unless it were in their interest, not merely in the interest of society as a whole, to do so. To convince them that it is in their interest, society must provide inducements that make it attractive for them to accept their unequally large share of risk.

To implement effective radwaste policy legislators might provide compensation to citizens or communities occupying the sites chosen for radwaste facilities. Compensation programs, whether offering increased community services like better schools or cultural programs or even making direct payments to individual citizens, might be effective for three reasons. First, the government must officially acknowledge—as it does during wartime—the sacrifice made by citizens in accepting risk for the benefit of society. Second, compensation programs must bring the appearance of voluntariness to the additional risks some citizens are being asked to bear. As the Expressed Preference Model makes clear, voluntarily assumed risks are more easily accepted than are mandatory ones. Third and most important, compensation programs accompanying radwaste policy decisions must bring the risks of nuclear waste disposal in line with the basic U.S. attitudes toward risk. Americans are imbued with the willingness to accept risk but only if it promises some reward. By providing incentives through compensation programs, the promise of reward would be present in radwaste policy, and federal policy for radwaste disposal would have a better chance of overcoming citizens' fears that have stymied any policy attempts.

CONCLUSION

As a byproduct of nuclear power and its attendant problems of radioactive waste disposal, the presence of risk calls attention to common social life in the United States. Risk is a predominant externality or social cost of many technological developments, and as such it calls for new policy initiatives aimed at overcoming the fear that risk engenders in people. In the United States with its liberal and capitalistic ideology, the fear is not necessarily of risk per se but of a lack of self-interested reasons for assuming additional risks. Radwaste policies that carried with them promises of reward for the acceptance of additional risks by the citizens most affected might provide those self-interested reasons and make both U.S. nuclear power and radwaste policies more effective. Without compensation schemes or other effective devices to engender public support, the future of nuclear power as a whole is very much in doubt.

Of course, compensation can be made to look rather unseemly—as an attempt to buy off the U.S. public through an appeal to narrow self-interest or simply individual greed, rather than through an appeal to patriotism or communal interests. But U.S. society is deeply individualistic, infused with belief in individual liberty and natural rights. These beliefs do not make social or communal life impossible, nor do they make social responsibility by citizens an impossible goal. What individualism does insist upon is the convergence of social with individual interests; it insists that individuals cannot be sacrificed to the common good without their consent. Policies like those relating to nuclear energy that have attempted to distribute risk have so far ignored these uniquely American sentiments, and, not surprisingly, have failed because of this omission. When the federal government begins to make policy for science and technology with these sentiments in mind, those policies will be more acceptable in ethical or constitutional terms.

NOTES

1. One of the most sensational cases of petrochemical waste disposal and the resulting environmental damage is that of Love Canal in New York State. For a thorough description of the events and problems at Love Canal, see the journalistic account by Michael Brown in *Laying Waste: The Poisoning of America by Toxic Chemicals* (New York: Pantheon, 1979).

2. *Science* 215, no. 4539 (March 19, 1982), pp. 1483–1486.

3. Quoted in *Science, ibid.,* p. 1484.

4. Quoted in *New Yorker,* October 1981, pp. 53–140; p. 54.

5. *Ibid.*

6. B. Kiernan et al., Legislative Research Commission, *Report of the Special Advisory Committee on Nuclear Waste Disposal,* no. 142, Legislative Research Commission, Frankfort, Kentucky, October 1977, p. 3.

7. *New Yorker, op. cit.,* p. 57.

8. *Ibid.*

9. *Science* 223, no. 4633 (January 20, 1984), p. 259.

10. K. S. Shrader-Frechette, *Nuclear Power and Public Policy* (Boston: D. Reidel, 1980), p. 51.

11. *New Yorker, op. cit.,* pp. 100–102.

12. *Ibid.,* p. 96.

13. U.S. Environmental Protection Agency, *Considerations of Environmental Protection Criteria for Radioactive Waste* (Washington, D.C.: EPA, February 1978).

14. *Ibid.,* p. 113.

15. *Science* 216, no. 4547 (May 14, 1982), p. 709.

16. *Ibid.*

17. Quoted in Shrader-Frechette, *op. cit.,* p. 49.

18. For examples of such scientific concern early in the 1970s, see Richard Lewis, "The Radioactive Salt Mine," *Bulletin of the Atomic Scientists,* June 1971; and Alvin Weinberg, "Social Institutions and Nuclear Energy," *Science* 177 (1972), pp. 32–34.

19. Lewis, *op. cit.*

20. Quoted in Dorothy Nelkin, *Controversy* (Beverly Hills, Calif.: Sage, 1979), p. 94.

21. Atomic Energy Commission, "Environmental Survey of the Uranium Fuel Cycle" (Washington, D.C.: AEC, April 1970).

22. *Science* 223, no. 4633 (January 20, 1984), pp. 258–260.

23. *Science* 219, no. 4580 (January 7, 1983), p. 34.

24. Nicholas Rescher, *Risk: A Philosophical Introduction to the Theory of Risk Evaluation and Management* (Washington, D.C.: University Press of America, 1983), p. 7.

25. Baruch Fischhoff et al., *Acceptable Risk* (Cambridge: Cambridge University Press, 1981), p. 27 passim.

26. *Ibid.,* chaps. 3–7.

27. See C. Starr, "Social Benefit versus Technological Risk," *Science* 165, no. 3899 (1969), pp. 1232–1238.

28. Quoted in Fischhoff, *op. cit.,* pp. 84–85.

29. P. M. Boffey, "Office of Technology Assessment: Bad Marks on Its First Report Card," *Science* 193 (1976), pp. 213–215.

SELECTED READINGS

Brown, Michael. *Laying Waste: The Poisoning of America by Toxic Chemicals.* New York: Pantheon, 1980.

Doniger, David D. *Law and Policy of Toxic Substances.* Baltimore: Johns Hopkins University Press, 1979.

Fischhoff, Baruch et al., *Acceptable Risk.* Cambridge: Cambridge University Press, 1981.

Lester, James P., and Ann O'M. Bowman. *The Politics of Hazardous Waste Management.* Durham, N.C.: Duke University Press, 1983.

Lowrance, William W. *Of Acceptable Risk: Science and the Determination of Safety.* Los Altos, Calif.: William Kaufmann, 1976.

Olson, M. C. *Unacceptable Risk.* New York: Bantam, 1976.

Rescher, Nicholas. *Risk.* Washington, D.C.: University Press of America, 1983.

Shrader-Frechette, K. S. *Risk Analysis and Scientific Method.* Dordrecht, Holland: D. Reidel, 1985.

Tomplin, Arthur R., and John W. Gofman. *Population Control Through Nuclear Pollution.* Chicago: Nelson-Hall, 1970.

6

COMMUNICATIONS TECHNOLOGY: POLICY FOR A WIRED NATION

Life was not going well for Dominic Suriano. Since the police of Columbus, Ohio, had come to his theater on March 14, 1980, and arrested him for showing pornographic films, his case in municipal court had been proceeding badly. On trial with three other defendants on charges of pandering obscenity, Suriano faced a conservative jury and a campaigning city attorney determined to resolve the issue of pornography in Columbus. Suriano's attorney Laurence Sturtz built his defense on a 1973 U.S. Supreme Court ruling that for a work of art to be judged pornographic, it must violate contemporary community standards.[1]

The community standards doctrine has been difficult to implement in law since its formulation because standards are not always readily apparent to either judges or juries. The doctrine has in effect given jurisdiction for deciding what is pornographic or obscene to states or localities; in most cases, including several in Columbus preceding the Suriano case, the courts have been reluctant to deem any material legally obscene. When obscenity charges result in conviction, they are commonly tried in conjunction with another crime that the prosecution argues is related to the pornography charge, such as rape or assault. In the Suriano case, the prosecution argued that the murder of an eight-year-old girl was the direct result of the accused murderer's viewing the allegedly obscene films Suriano had shown in his theater. The prosecution's arguments echoed those of a Baptist minister, who remarked after viewing the two films in question, "it's no wonder [she was killed]; this feeds their lust."[2]

Given the ambiguity of the community standards doctrine and the increasingly dire predicament of his client, Larry Sturtz attempted a novel legal gambit. Arguing that the community had already passed judgment on the nature of the films in question, he subpoenaed the subscription records of two cable television companies operating in the Columbus area that had shown the films as part of their adult programming package. The companies, Warner Qube and Coax Cable, resisted the subpoena on the grounds that it would compromise their subscribers' rights of privacy. Sturtz argued, however, that home viewing of the films by a large number of citizens would constitute public acceptance of them as within community standards of decency; therefore, subscription information was essential to the defense of his client. Sturtz won permission to subpoena the subscription records in county court, and after some delay both companies complied by supplying raw subscription figures without names. The figures showed that over ten thousand Columbus residents had, in the comfort of their own homes, viewed "Taxi Girls" and "Captain Lust."

Dominic Suriano was acquitted of pandering obscenity. In the wake of his courtroom victory were thousands of citizens clamoring for assurances that their privacy would be protected by the cable industry and at least two major cable companies struggling to provide reassurance to their customers.

DEVELOPMENT OF CABLE COMMUNICATIONS

Cable technology includes an array of new and developing techniques and devices that together make up what is often referred to as the "communications revolution." These technologies include satellite communications, microwave and narrow-beam transmission, computer networking (worldwide computer hookups), and light beam telephone transmission. Together these technologies constitute an almost bewildering set of possibilities for communication in what cable consultant and futurist Ralph Lee Smith has called the "Wired Nation."[3] Each of these technologies represents a major step in communication capabilities for government, industry, and private citizens. They also raise legal, constitutional, and policy questions at least as serious as the right of privacy issue already seen in the Columbus obscenity trial.

In this chapter we will focus on cable technology. In many ways this technology offers the most accessible and varied group of new communication opportunities to the largest group of citizens—roughly to everyone in the United States who owns a television set. Through a cable hookup, viewers are supplied clear and crisp television pictures and a selection of network and local stations. Cable services do not stop there. Depending on the sophistication of the cable system, owners can watch a variety of news and entertainment subscription stations, control their energy usage, receive immediate Wall Street information, shop at home, bank at home, read library books through their television, tie into large library and information systems including the Library of Congress, and vote in town council meetings while watching the proceedings from their living rooms. Although all these options are currently offered by only the most advanced cable systems, like those in Columbus and Cincinnati, Ohio, franchises for cable that include such services have already been awarded in several other U.S. cities.

As Smith pointed out, as recently as 1970 all this cable largess was merely speculation and was called by the *Yale Review of Law and Policy* the "cable fable."[4] Cable television had been in existence since the early 1950s, when it was installed mostly in rural areas where over-air television signals were weak because of the long distances from transmitters. However, by the 1960s some academics and researchers began to realize that cable could do more than simply deliver clear pictures of Andy Griffith and Milton Berle. Cable's prospects for community access television and interactive communication were widely discussed; glowing predictions were made concerning communities brought together to participate in "electronic democracy." But cable television (CATV) did not catch on in the 1960s: It was too expensive to provide in rural areas, and city dwellers had few problems with station reception.

In 1975, one man and one decision changed the nature of cable television and started it on its way to becoming a boom industry. Chuck Dolan owned a fledgling pay television service named Home Box Office (HBO), which for a monthly fee showed movies and adult programming to cable owners in New York City. His employers at Time, Inc., owned Sterling Manhattan Cable that served all Long Island and carried Home Box Office as an option. In 1975, Dolan convinced Time to purchase time on RCA's communication satellite (Satcom I) in order to beam HBO to cable owners anywhere in the country. For a demonstration, Dolan invited industry executives to view on a large screen a championship boxing match live from the Philippines via the satellite. The picture was so sharp and clear that HBO was a sensation, and the rush was on to connect cable television to the satellite. Within a year 90 percent of Sterling's Long Island subscribers were paying extra for HBO. By 1976, the franchising race was under way, and cable television was a financial gusher.

Still, not until the late 1970s did cable technology catch up with the financial commitments of its backers. When Warner Amex installed the new QUBE system in Columbus, Ohio, the potential of cable television as a total communications system in the home was finally realized. And, as often happens with rapidly developing new technologies, this was also the time when the legal, constitutional, and political questions concerning cable communications finally—and belatedly—were asked.

CABLE AND THE LAW

As an industry, cable television must answer to many legal and regulatory masters, and as a result it is often subject to none. As contradictory as this sounds, the multitude of federal, state, and local agencies that have legal jurisdiction over cable television ensures that cable is one of the least regulated technologies that we discuss in this book. As one source pointed out, "from the beginning, local governments granted franchises, licenses, and permits to community antenna television systems, although many systems emerged without any government authorization or attention."[5] Indeed, granting franchises to cable companies is the stage at which many of the most serious political and legal questions concerning the technology arise. These questions have not yet been answered because of the political pressures and recurring problems in intergovernmental relations.

Since 1968, the Federal Communications Commission (FCC) has been the ostensible regulatory authority for cable systems. This authority was not assumed without challenge. In fact, the FCC itself had ruled in 1959 that it had no authority over the cable systems that existed at that time. However, the Supreme Court mandated FCC authority in 1968, and in that year the FCC developed a comprehensive set of cable rules.[6] The rules went into effect in 1972 but were so complicated and detailed that cable companies, the public, and even the FCC had difficulty comprehending them.[7] Thus, later in that year the FCC issued a reconsideration and a clarification, both of which took considerable authority away from the FCC and gave it jointly to states and local governments.

Federal and local sharing of regulatory authority over cable television has frequently led to confusion over what states, cities and towns, or individual citizens can demand from cable companies. Since 1972 the FCC has reserved certain preemptive authority over some aspects of cable construction and delivery, but even in these areas states or localities can assert authority if they can prove that they have "the authority to consider and do consider the interests of subscribers of cable television services, as well as the interests of the consumer of the utility services."[8] The areas of preempted authority are (1) technical standards; (2) local origination, which grants to cable systems immunity from local authorities in determining what programming is carried on cable; (3) signal carriage—the FCC determines which stations must be carried on each cable system; and (4) subscription rates and franchise fees.[9] Since 1980, the FCC has relinquished authority over many of these areas and turned it over to the states and local governments, largely because of political pressure from the Reagan administration.

Who regulates what in the cable industry remains an unanswered question for all involved in cable television from the FCC to the home viewer. This uncertainty is manifested in a number of perplexing possibilities and legal questions surrounding cable, which arise at every level of cable construction and operation, and must be asked if only to determine who or what has the authority to provide answers. We now turn to the most interesting and puzzling of these questions.

THE FRANCHISE WARS

Since 1976, the boom in the number of franchise applications submitted by cable companies hoping to be awarded operating and service rights has had all the frenzy of a nineteenth-century gold rush. And with reason: The profits anticipated to accompany every franchise awarded are immense. It is estimated that franchises built even since 1980 can be sold for up to 150 percent of their original cost.[10] Competition for franchise areas is so frantic that it has forced some companies to make outlandish promises and shady political deals that are an embarrassment not only to the industry but often to public officials as well.[11]

Private Ownership

Part of the urgency in the competition for franchises arises from the nature of the technology itself. Running cable is an increasingly costly proposition, particularly for remaining franchise areas that encompass rural districts with low population density. It may cost as much as several thousand dollars for a cable company to wire one household for cable if that household is located well outside of town in a sparsely populated area. And after a month of service that customer might cancel his or her cable subscription, thereby costing the company almost all its initial investment. This example is extreme, but the point is clear: The investment risks for a cable operator are very high, and potential profits usually lie in the not so near future.

Because of the high initial costs to a cable operator, in practice franchises are almost always awarded on a monopoly basis, though legally franchises are termed nonexclusive. Local governments in most states have the home rule authority to make the franchising decision, usually in conjunction with the state department of public utility regulation. Once the franchise is awarded to a cable company, no other operator is allowed to overbuild in the same region. This exclusivity obviously makes getting the franchise a matter of great contention among companies. In the competition issues like universal service for all residents in the region, free wiring or low installation costs, and number of channels become the arenas for sincere promises by competing companies, some of which cannot possibly be kept after the award is given. As one commentator stated, "It is highly probable that many cable companies will default on commitments or be delayed in their delivery on franchise promises while capital and the technology catch up. Cities now in the franchising process or working with the supersystems now under construction will have to recognize the sugar plums for what they are."[12]

Difficulty in keeping promises is nothing new in U.S. political decision making, but the behavior of cable companies in the franchising stage, as well as the prospects of future profits, has moved some local governments to rethink the question of cable ownership and operation. Most franchises are owned by private cable companies, but the number of exceptions is growing. In St. Paul, Minnesota, and Grosse Point, Michigan, cable services are owned and operated directly by local government agencies. The possibilities do not stop there: Cable systems may be owned by the city but operated by a private company; owned and operated by a public, nonprofit corporation, cooperative, or power company; or owned jointly by the franchising authority and a private company under a variety of sharing arrangements.[13]

Public Ownership

When cable services are publicly owned, some sticky legal and constitutional questions arise. First, it is not altogether clear whether entertainment should be provided to private citizens by a public authority. Questions of censorship may arise in such a situation, although this question of "local origination" of cable services is somewhat broader in the context of private ownership.

More significant is the right of privacy issue. Although this right is nowhere mentioned in the U.S. Constitution, the Supreme Court has frequently reiterated its status as a penumbral doctrine, meaning that the Bill of Rights implies that the right of privacy pervades all other rights listed in the first ten amendments. When cable services include, as they do in many franchise areas, opportunities to shop and bank at home, the role of public authority is potentially a dangerous one. Would bank statements from cable banking transactions be made available to local taxing authorities or to the Internal Revenue Service? Would energy monitoring via cable be subject to public review in the case of energy emergencies? And if the cable services provide the capacity to watch town council meetings and to be polled at home concerning issues, would this service constitute political discrimination if not offered to all citizens? And even if it were offered to all citizens, would monetary charges for the service constitute a modern equivalent of the poll tax, now considered unconstitutional? Finally, given the cost of operating a cable system, is the record of public ownership and operation of any essential or valued service sufficiently illustrious to expect maximum efficiency from a publicly owned and operated system?

A recent federal court case in Boulder, Colorado, is relevant to the issue of the local, public ownership of a cable franchise.[14] A cable operator named Cable Communications Company was denied a franchise and brought suit against the City of Boulder on the grounds that the municipality was acting in violation of antitrust laws. The court accepted this novel concept and voided the franchising decision made by the local government agency. Eventually upheld by the Supreme Court, the decision sent shock waves not only through the cable industry but also through virtually every city and town hall in the country. If city and state authorities could not in fact award franchises or contracts on the basis of competitive bids, most of what passes for local politics and home rule authority would be lost. The issue is likely to be raised in court again, but in the interval local politics has been radically affected by the legal interpretations surrounding a relatively new technology.

Local Origination Versus Common Carriership

Intimately related to the question concerning ownership of cable services is the question of the nature of the service provided by the operator. This complicated issue involves the rights of the cable company to determine the substance of the programming available to its subscribers. Cable companies claim the right of local origination—they reserve all authority to decide the type of programs and content offered to their viewers. This authority gives the operators the status of (in their terms) electronic newspapers, with the companies themselves serving as sole editors of programming content.

Opposed to the concept of local origination is that of common carriership. If a communication technology or company is classified as a common carrier, it can legally provide communication capability but has no authority over the material actually communicated. The best example of a common carrier is the telephone industry, and the Bell System in particular. The phone company provides the service but has no power to regulate what is actually said in any telephone conversation.[15] If Congress designated cable companies as common carriers, they would lose control over what could actually be seen on cable television. This power would then be transferred either to the FCC, to local and state agencies, or to the citizens themselves.

Losing the right of local origination would clearly be catastrophic to many cable companies seeking new franchises, for in most cases the different programming alternatives that the companies offer set them apart from each other and give each a chance of winning the franchise. Cable companies argue for local origination and against common carriership on the basis of First Amendment guarantees of freedom of speech. However, this high-sounding argument cuts both ways: If cable companies are given carte blanche in determining who can broadcast over the cable, they can just as easily deny access to any group or entertainment organization. Freedom of speech for cable operators can be tantamount to censorship of a different kind: censorship carried out by cable companies with local origination authority and exercised against groups seeking to exercise their right of free speech by using the cable system.

The possibility of industry censorship is particularly troubling on public-access stations, which almost all cable operators include in the franchise package. These stations are run by the cable operator for the use and benefit of the community, and air time is presumably filled on a first-come, first-served basis. However, discrimination against some community groups does occur on public-access stations, and even on regular cable stations programming of special interest to minority groups has been woefully lacking.

But the issue most vociferously raised regarding the content of cable offerings involves pornography. As in the Columbus case, explicitly sexual or violent material is readily available on many cable systems, even though a majority of community residents may not watch it or even want it as part of the franchise package. This controversial material raises a different issue of censorship, of course— one that remains unresolved either by the courts or by the state and federal cable-regulating authorities.

In fact, little regulatory action by either states or the FCC has been upheld in court since 1978. According to rules adopted in 1972, the FCC required cable operators to meet traditional guidelines for television regarding the airing of obscenity or material depicting explicitly sexual or violent acts. Also, the FCC required every cable operator to provide a certain number of public-access stations as part of its franchise package, depending on the number of channels it was offering overall. In 1978, a federal judge struck down these guidelines on the grounds that no adequate basis existed for them under the Communications Act of 1934.[16] Thus, important regulatory powers have eroded both from states and from federal authorities like the FCC.

With the spirit of deregulation now gaining preeminence in the franchising of cable technology, cable operators are enjoying powers explicitly denied to motion picture companies in the late 1940s. In the early days of the film industry, major movie studios also owned large chains of local theaters. With the powers of production, exhibition, and distribution that this vertical integration provided, major studios were able to control which films became hits by not showing films from competing studios in their theaters. In 1948 the Justice Department issued a consent decree forcing studio owners to divest themselves of their theaters on the grounds that these monopoly arrangements violated the freedom of choice of film watchers as well as the economic freedom of independent theater owners. Now, through the local origination authority of cable companies—several of which produce their own entertainment programs—"the centralized control of programming, distribution, and exhibition is sneaking back into the marketplace."[17]

CABLE AND CONSTITUTIONAL CONUNDRUMS

Like many developing technologies, cable communication has given rise to several constitutional challenges, some of which have been referred to in this chapter. Issues of local origination, two-way interactive services (e.g., banking, shopping, or polling), and privacy have all involved questions regarding constitutional guarantees. So far few of these questions have been resolved for either society as a whole or even the entire industry. Some individual cable companies like Warner Amex have on their own initiative adopted codes of conduct that include so-called privacy codes. But, though laudable, these actions do little to resolve issues for the entire industry and usually extend only to the franchise area open for bid.

Discrimination in Dissemination and Programming

During the franchising stage of cable deployment, potential customers often raise charges of unconstitutional discrimination. Because of the expense and financial risk involved in laying cable, particularly in rural areas, regions with low population density are the last to be approached by cable operators for franchise rights. Indeed, some areas of the country may never be wired. Does this fact constitute undue discrimination against farmers, small-town dwellers, or people who prefer the slow pace of rural life but still demand cable service? Or is the lack of cable television simply part of the equation when individuals decide where they want to live? If cable technology involved only increased entertainment options these questions could be easily answered. But given the considerable communication benefits that cable offers beyond entertainment, deprivation of the chance to receive cable service might be compared to deprivation of the option of telephone service.

In the last five years of cable franchising competition, other accusations of discrimination have been leveled against the industry. Even in densely populated urban areas, some neighborhoods have had difficulty finding cable operators willing to bid for franchise rights. Inhabitants of poor areas of cities in which low-income households predominate, discover that cable comes late to their neighborhoods. The results of studies by cable companies and consultants discourage franchise competition there, because poor families do not buy additional programming like HBO or interactive services for which profit margins are high.

Also, racially and ethnically heterogeneous areas experience difficulty in acquiring cable because of demands for several public-access stations to accommodate each different ethnic group. Because the FCC no longer requires a specified number of public-access stations, companies are reluctant to provide many of them. Therefore, if these areas do receive cable bids, they frequently find that none of the bids offers the full range of public-access services.

After franchises are awarded the constitutional dilemmas are not automatically resolved. Once programming begins, members of some minorities have claimed that discrimination continues against them through the kind of programs delivered by the operator. The absence of blacks, women, and other minorities in entertainment roles that do not rely on sexual or ethnic stereotypes is a recurring complaint and, of course, is not unique to cable programming. However, because cable is a pay service, some customers insist that the obligation of operators to provide nondiscriminatory, nonstereotypic images of minorities is increased. Also, because few advertisers buy commercial time on cable systems, those who feel insulted by cable programming do not have the option of writing to advertisers to complain about programs or to threaten boycotts of products unless programming improves.

Electronic Democracy

By far the most perplexing and interesting political challenge presented by cable is that involving the implications of interactive or talk-back capability. As previously mentioned, the idea of electronic democracy intrigued some academic researchers in the 1950s, but in places where this capability actually exists, it is regarded with considerably less optimism. In Columbus, Ohio, viewers have the option of watching ongoing town council meetings and being polled about their opinions on certain questions. But what exactly is the status of their answers? Do responses indicate the actual wishes of the citizenry on an issue like a new zoning ordinance, or should members of the council ignore the results of the poll? And who is doing the voting at home—a responsible adult voter or an eight-year-old child who has recently discovered the new buttons on the television? If indeed the council is assured that the poll results are legitimate, how should the poll affect the vote by the council itself? And which vote has the more binding force—the formal council vote or the television poll?

Although these questions are clearly not rhetorical, they are not easily answered. Cable technology, particularly the Warner Qube system, has reintroduced the idea of participatory democracy, but it is not clear whether this development is beneficial to U.S. politics. Certainly state and local political figures do not necessarily view television polling as an improvement in their ability to govern.[18] They worry—with reason—about what will become of U.S. political institutions if interactive capability is taken too far.

As a representative democracy, the U.S. political system is not fashioned to function in an explicitly direct, participatory way. Institutions from the Congress to the town councils are predicated on the idea that the wishes of voters are filtered through various representative devices, instead of made immediately into the law of the land. In other words, representative bodies are expected to deliberate on the need for and purpose of new laws, and their role is to provide that deliberation and balance in political decision making that individual citizens are either incapable of providing or unwilling to do themselves. Ideally, in the deliberation all interests would be considered, including the good for society as a whole, rather than merely for certain individuals or groups.

No system is perfect, of course, and certainly many interests go unrepresented or their views unarticulated in U.S. politics. But is the type of participatory democracy promised by interactive cable systems an improvement on politics in the United States? If cable offers the prospect of increased public participation, the question should be asked whether as a society the United States is indeed ready to take that leap into participatory democracy. Are citizens informed or interested enough in the affairs of state (or even town) to warrant increased reliance on television polls in making political decisions? Would even enough citizens watch the proceedings in the town hall (or as a future scenario—the Congress) to warrant reliance on the results of the poll? And even if they did watch in sufficient numbers and use their televisions to respond, what would be the long-term results? It is questionable whether U.S. politics either needs or wants such increased participation, and given the usual workings of legislative bodies at all levels, this participation might lead to increased disillusionment and alienation of citizens who see for the first time how Congress or their town council really works.

Furthermore, it is worth asking whether this type of participation is good for U.S. politics. Sitting in the privacy of one's own home and making important political decisions is far different than the public give and take of open political forums. It is arguable that cable provides a political experience that is isolating and therefore not conducive to political decision making in the truly public interest. Political theorist Jean Bethke Elshtain echoed this concern.

> The ersatz participation characteristic of interactive television is dramatically at odds with this democratic ideal. . . . Watching television is an isolating experience. . . . Television is privatizing: it appeals to us as private consumers, not as public citizens. . . . [I]nteractive systems encourage social atomization and they foster the notion that an electronic transaction is an authentic democratic choice. That so many people see democracy alive and well in electronic beeps, flashes and commands to "register your opinion now" shows how confused we are about the essential nature of choice. To see button-pressing as a choice, as a meaningful act on a par with marching in an antinuclear rally, lobbying against toxic waste dumping or working for a political candidate, indicates our tacit embrace of a crude version of the "preference theory" of economics.[19]

Certainly the acceptance of push-button democracy would dramatically alter the nature of the U.S. political system in other aspects as well. The reliance by legislators at every level on lobbyists and interest groups as sources of information as well as representatives of salient political issues would greatly decline. This development may not be seen as entirely negative; yet interest groups serve important

and democratic purposes in the political system as presently constituted, and their function as information providers could not be filled
as well by simple push-button polling.

Finally, all this public airing of issues and polling through the
home television set may have consequences that are unforeseen and
opposite of what is intended. Video polling may conceivably make
government more secretive in its operation rather than less so. When
the legislative debate over issues becomes instantly available to the
voter at home, the real power in politics would certainly be that of
deciding what those issues are and when they will be subjected to
"public" debate. If the use of electronic democracy were to become
widespread, the present power of the citizen in this area (such as it
is) would most likely decrease. The ability to set the political agenda
is always an important source of power in a democratic political
system. Too much emphasis on the public's right and new-found
capability to "vote" on issues might obscure its rightful and more
significant power to decide what those issues are.

Computer Crime

One last area of constitutional quandaries introduced by cable
technology should be considered: the field of crime. Many recent
news stories have reported the rising incidence of computer crime
by clever computer-assisted felons and by relatively innocent and
often youthful computer hackers. The capabilities of cable are disturbing in this regard because most franchises offer special services
to customers who own home or micro computers. Aficionados can
gain access to data banks around the country and even the world
simply by purchasing time on cable-provided computer networks.
The consequences of this opportunity are not yet known but they
warrant particular attention.

In some legal areas an interesting result of the introduction of
cable has been the difficulties surrounding the very definition of
crime. Theft of cable services has become the plague of the industry;
yet defining what exactly constitutes the criminality of some forms
of theft has proved almost impossible. Splicing into someone else's
cable in order to wire one's own set is clearly a criminal act and
punishable with a fine or even a prison term. But is stealing a
television signal directly from a satellite by means of a personal dish
antenna also theft? Most cable pay services like HBO beam signals
directly from satellites to ground receiving stations and then send
them out over the cable. But the signals can be intercepted with ease
directly from the satellite by increasingly inexpensive rooftop or

backyard dish antennas. As yet the courts have not declared this interception illegal, though cable operators estimate that they lose millions of dollars annually to these ham operators. Some companies have decided to scramble the message at the satellite and provide descramblers either at receiving ground stations or in the home; however, the companies have found that descramblers are easy to construct and sell.

The legal issues of such activities are complicated because they involve questions concerning ownership of the airwaves. Also, because communications satellites are deemed common carriers, regulation of transmission is constitutionally questionable on privacy grounds, as in the case of telephone companies. Thus, though some guilt-ridden antenna owners have even sent HBO monthly checks to pay for programs intercepted from the satellite, until satellite interception is actually defined as a crime, the cable companies must return the checks and lose the profits.

CABLE AND THE FUTURE

By its very nature cable seems a futuristic technology. Yet in many parts of the country this future has already arrived. Although it is tempting to predict the unforeseen benefits cable technology may someday provide, the time for unwarranted promises and uninformed speculation has passed. Indeed, events in the cable industry have already shown how difficult prediction is in the whole field of communications. Recent rulings by the FCC encouraging microwave television construction in urban areas have somewhat dampened the cable industry's prospects. Also, in many subscribers' eyes programming on some cable stations is not worth the price: Some pay stations merely show old movies and reruns of antiquated network programs. Most industry analysts agree that the programming software of cable has yet to catch up with the enormous advancements in hardware. Until software does catch up, cable operators will have a difficult time getting new customers or even keeping present ones.

Yet the almost wondrous potential of cable communication cannot be denied. Unfortunately, the same is true for the less wondrous but equally surprising policy problems to which cable has given rise. As a new technology in need of relevant policy decisions, cable is very instructive in terms of the general approach to science and technology policy. We have seen in previous chapters that the formulation of sound policy for science and technology needs increased citizen involvement from the very start of the policy process. Such participation can be defended on the grounds that it is necessary because of the rightful dictates of a democratic system and also because examples abound of the dire results of policies made with inadequate citizen input.

Because cable policy affects citizens where they are very sensitive—in their own homes—they should naturally take part in making policy for this new technology. In most towns citizens are already active in the franchising stage: Citizen groups are either appointed or informally established to advise in the awarding of the contract. This input is not only admirable but advisable in other areas of decision making concerning the course of cable development. Furthermore, making policy for cable technology is a good training ground for policy makers at all levels, from the FCC to the citizen who watches television. The intricacies of cable policy are unique and difficult, but they resemble those of other technologies we have examined in this book. How well citizens and their government establish policy in this area may well determine the course of communications in the coming decades, both between citizens and between citizens and their leaders.

NOTES

1. *Columbus Dispatch* 109, no. 315 (May 10, 1980), p. 1.

2. *Ibid.* See also *Columbus Dispatch,* May 28, 1980, p. B-10; June 1, 1980, p. B-7; June 11, 1980, p. B-6; June 12, 1980, p. C-1; June 13, 1980, p. B-1.

3. Ralph Lee Smith, *The Wired Nation* (New York: Harper and Row, 1972).

4. Ralph Lee Smith, "The Birth of a Wired Nation," *Channels* 1, no. 1 (April/May 1981), p. 33.

5. Thomas Baldwin and D. Stevens McVoy, *Cable Communication* (Englewood Cliffs, N.J.: Prentice Hall, 1983), p. 168.

6. *United States* v. *Southwestern Cable Co.,* 392 U.S. 157, 1968.

7. Baldwin and McVoy, *op. cit.,* p. 169.

8. *Ibid.,* pp. 171–172.

9. *Ibid.,* p. 174.

10. *Ibid.,* p. 191.

11. In a recent example in Connecticut during a franchising competition in a relatively small region of the state, a prominent political figure was made a partner in the cable company that hoped to win the franchise and was given special stock considerations if the company won the contract. Though the company arguably offered the best service, it was denied the franchise when news of the agreement—which was perfectly legal—was reported in the media.

12. Baldwin and McVoy, *op. cit.,* p. 192.

13. *Ibid.,* p. 189.

14. *Community Communications Co.* v. *City of Boulder, et al.,* 80-1350, January 13, 1982.

15. Except in the case of obscene phone calls, of course. But even here authority derives not from the rights of the telephone industry but from law enforcement agencies.

16. Ralph Lee Smith, "The Birth of a Wired Nation," *op. cit.,* p. 88.

17. *Ibid.*

18. The experience of the authors bear this out. In performing community needs ascertainment studies in northeastern Connecticut as cable consultants, we were frequently told by local officials and political leaders that "we don't want that interactive polling to be part of the franchise package."

19. Jean Bethke Elshtain, "Democracy and the Qube Tube," *The Nation,* August 7, 1982, pp. 108–110.

SELECTED READINGS

Baldwin, Thomas, and D. Stevens McVoy. *Cable Communications.* Englewood Cliffs, N.J.: Prentice Hall, 1983.

Compaine, Benjamin M., ed. *Who Owns the Media: Concentration of Ownership in the Mass Communications Industry.* White Plains, N.Y.: Knowledge Industry Publications, 1972.

Mahony, Sheila, et al. *Keeping Pace with the New Television: Public Television and Changing Technology.* New York: Carnegie Corporation of New York, VNY Books International, 1980.

Muth, Thomas A. *State Interest in Cable Communications.* New York: Arno Press, 1979.

Smith, Ralph Lee. *The Wired Nation.* New York: Harper and Row, 1972.

7

RECOMBINING GENES:
SCIENTIFIC LIBERTY
AND PUBLIC SAFETY

Even the most steadfast faith in the fruits of scientific endeavor can be shaken by the impact of discovery. At times scientists may sense that exploring the frontiers of science can carry an excitement that supplants an innate and necessary caution in research and can lead to a disregard of the wider concerns of public interest and safety. Scientists in the early 1970s concluded that genetic research was an area of investigation so thrilling, yet so fraught with potential dangers to the public, that because of their responsibility to both science and society they had to pause in their research and consider the public's stake in their research. The result was an example of self-regulation in science remarkable for its public spiritedness yet troubling for its naiveté and disregard for the processes of public policy making in the United States.

The story began at a little-publicized and sparsely attended scientific conference in New Hampshire in summer 1973. The attention of the news media was focused elsewhere, as the first congressional Watergate hearings convened in Washington, D.C. During the final evening of the Gordon Conference, a panel discussion headed by Maxine Singer of the National Institutes of Health (NIH) and Dieter Soll of Yale considered the potential hazards of research currently being done with animal genetic material, which involved the splicing and re-combining of DNA molecules and resulted in new forms of human-created viruses. Attending scientists decided to write letters to the president of the National Academy of Sciences (NAS) and to the weekly scientific journal, *Science,* expressing concern over the possible exposure of the public to new viral strains created in the laboratory.

Considerable concern had already been expressed about recombinant techniques for DNA, but at the conference a new and easy method for recombining DNA developed by Paul Berg of Stanford was discussed at one of the technical sessions. This technique raised to new heights the scientists' uncertainty over the potential health hazards presented by widespread recombinant research, though the attitude of the public and press did not yet mirror that concern. This professional uneasiness was expressed in a *Science* article in November 1973, in which eminent scientist Robert Pollack stated that "we're in a pre-Hiroshima situation. It would be a real disaster if one of the agents now being handled in research should in fact be a real human cancer agent." Similarly, prominent geneticist Wallace Rowe admitted that "the Berg experiment scares the pants off a lot of people, including him."[1]

In response to the letter sent by the participants of the Gordon Conference, NAS president Philip Handler appointed a blue ribbon panel of scientists headed by Paul Berg to study the need for tighter controls on recombinant research. The work of the panel resulted in the now famous "Berg letter" written to the editors of *Science* and the British science weekly *Nature,* calling for a moratorium on recombinant research until more effective safety measures could be developed.[2] The daily press around the world picked up the story, and pressures mounted on the scientists engaged in recombinant research and on government agencies to provide safety guidelines.[3]

As a result of the Berg letter, the NAS announced an international meeting to be held at the Asilomar Conference Center in California in February 1975 to draw up new measures for laboratory safety and containment in recombinant DNA experiments. Only scientists currently engaged in recombinant research were invited to attend the conference; over the objections of a small but vocal minority, a report recommending a series of new safety procedures was approved and sent to NAS. In James D. Watson and John Tooze's description of the proceedings, Paul Berg showed a mastery of the political tactic of obfuscation in voting as he presided over the report's passage. Nevertheless, over the objections of Watson and others, "the majority at Asilomar either declared, or failed vocally to deny, that special guidelines were necessary; they at least agreed that there were valid reasons for concern, that there might be significant hazards."[4] The newly endorsed safety precautions ranged from the most elementary strictures on eating and smoking in the laboratory or pipetting by mouth to the very sophisticated and expensive methods of physical and biological containment of genetic material.[5]

By most accounts, the Asilomar conference was a noble effort and at least a partial success.[6] According to Watson and Tooze, however, the expectations of scientists at the conference were unrealistic at best. The participants assumed that by regulating themselves, they would thereby be free from external control or pressure by either governmental agencies or a concerned public. "Having demonstrated their integrity," Watson and Tooze reported, "they naively believed that they would now be free of outside intervention, supervision, and bureaucracy."[7] This freedom did not materialize, however: Both regulation and public controversy over DNA research were just beginning.

Conspicuous in their absence at the conference were several groups of scientists and environmentalists familiar with recombinant research and opposed to it. Their absence was itself a point of controversy in the later development of policy disputes over the research, although formal environmental groups like Friends of the Earth had not yet considered the issue. More significant was the absence of a group called Science for the People, made up of eminent molecular biologists, including George Wald, Ruth Hubbard, and Erwin Chargaff. These biologists had themselves conducted recombinant DNA experiments and were convinced that the research should be stopped: (1) It was too dangerous in terms of public safety; (2) it constituted an unethical intrusion into the natural evolutionary process. Wald, a Nobel laureate in biology, later expressed his objections in the Cambridge, Massachusetts, city council hearings, which were called to consider whether recombinant research should be banned within city limits. The Asilomar conference, it turned out, was merely the first step in the development of regulatory policy aimed at protecting the freedom of inquiry of scientists while securing the public from undue risk.

REGULATORY TANGLE: NIH AND CAMBRIDGE

Development of Safety Regulations at the NIH

Even before the Asilomar Conference had convened, but unbeknownst to its participants, the National Institutes of Health had taken steps to begin regulation of recombinant research. The Berg letter to the NAS had recommended that the NIH establish an

advisory committee charged with (i) overseeing an experimental program to evaluate the potential biological ecological hazards of the above types of recombinant DNA molecules; (ii) developing procedures which minimize the spread of such molecules within human and other populations; and (iii) devising guidelines to be followed by investigators working with potentially hazardous recombinant DNA molecules.[8]

Within three months after the letter appeared in *Science,* NIH director, Robert S. Stone, established the Recombinant DNA Molecule Program Advisory Committee (RAC). Though the committee ostensibly carried only advisory status, it was charged with "devising guidelines to be followed by investigators working with potentially hazardous recombinants." After Asilomar, the RAC became almost totally concerned with this aspect of its charter.[9]

The NIH is a research and administrative agency, not a regulatory one, and none of its committees has regulatory authority. The lack of power to enforce safety guidelines became a major drawback to the RAC's ability to actually generate research policy, though this development did not really appear until 1983. Before that time, all scientists initiating research involving recombinant experiments voluntarily submitted their proposals to the RAC for approval; submission was required only for those projects relying on NIH funds. Recombinant research carried out in private laboratories, particularly in corporations developing new products using recombinant methods, was not officially subject to the RAC's advisory approval. This division would cause problems down the road in the development and exploitation of the technology.

Drawing Up Guidelines. The RAC drew up guidelines in spring and summer 1975, and they were approved in December of that year. The committee was directed by two assumptions: first, that it should attempt to provide encyclopedic guidelines that would cover all possible classes of recombinant DNA experiments; and second, that the potential public safety risk of exposure to dangerous new viruses arose as the phylogenetic and evolutionary makeup of the material whose DNA was being manipulated grew closer to that of the human species. In other words, if molecules used in research were similar to those in the human body or actually present in it, the risk of exposure to humans was greatly increased. Because in 1975 most recombinant research used molecules of *Escherichia coli (E. coli),* the risk was considered very high, as this bacteria is constantly present in the human bowel. The high risk of using *E. coli* proved important in the development of guidelines because it moved the committee to recommend biological as well as physical containment measures to ensure safety. Biological containment would entail the use of special strains of *E. coli* that could not live outside the laboratory.[10]

The RAC had difficulty drawing up guidelines because the scientific community was involved in the writing of the proposed guidelines and because the scientists themselves were motivated by two opposing objectives. On one hand, some scientists (for instance, Paul Berg) insisted on strict guidelines to minimize the hazards of the research, even if that meant that certain experiments would be banned. Others (like George Wald and Erwin Chargaff) argued that no experiments involving any type of *E. coli* bacillus should be allowed. On the other hand, microbiologist James Watson argued that any guidelines regarding *E. coli* would be too restrictive and would effectively close off research in the most promising areas that required the use of DNA from warm-blooded animals and from viruses. Watson insisted that this research—though admittedly somewhat more dangerous— offered the greatest opportunities for human betterment in "developing bacteria as new sources for the production from inserted genes of human proteins such as interferon, insulin, and growth hormone."[11]

The committee's work was hampered by the openness of its deliberations. The first set of guidelines, proposed by a subcommittee in June 1975, was discussed and rejected at a scientific conference the next month as too stringent. Subsequently loosened, the guidelines were then vehemently criticized as too lax by Berg and other biologists; they subsequently submitted a petition to the RAC calling for more strict and thorough controls. A new subcommittee, formed in response to the criticism, presented stiffer guidelines (also criticized as being too harsh) that the RAC and NIH eventually adopted in December. Despite the acceptance of the guidelines by the RAC and NIH, in the end these agencies had no power to enforce them for research not conducted with NIH funding.

Safety Regulations. The safety regulations accepted by the RAC at the end of 1975 were a complicated and inclusive collection of physical and biological containment measures. The physical containment guidelines identified four levels, from P1 to P4, in ascending degree of strictness, depending on the nature of the experiments. For actual recombination of DNA molecules, only two levels applied, P3 and P4. Briefly, the physical containment requirements were as follows:

P1: Use standard microbiological techniques.
P2: Same as P1, but hang a Keep Out notice on the door while the experiment is in progress.
P3: Same as P2, but put the lab under less than outside air pressure so that if a door were accidentally opened air would rush into the lab instead of out, or if you cannot manage that, at least use negative pressure cabinets.

P4: Same as for handling really dangerous agents (such as anthrax bacilli, smallpox viruses, and the like)—use air locks, negative pressure, change clothes and shower, and so forth.[12]

The biological containment techniques began with the assumption that *E. coli* bacilli found in the human body and other warm-blooded animals as hosts of recombined DNA molecules were too dangerous to use. Therefore, the minimum level (among three levels) of biological containment (EK1) would require the use of *E. coli* K12, a variety of the bacillus that can survive only under special laboratory conditions and can neither survive nor reproduce in the human body. Briefly, the levels of biological containment were as follows:

EK1: Use the standard laboratory version of *E. coli* K12 as the host for your recombinant DNA molecule, and use *E. coli*'s standard virus or plasmids (independently replicating bacterial chromosomes) as the vector (host).

EK2: Use strains of *E. coli* genetically altered so as to be *in theory* 1 million times less likely to escape successfully from the lab than standard *E. coli* K12.

EK3: Same as EK2, except that someone has gotten around to actually confirming by empirical test that the disarmed bug is indeed 1 million times less likely to make a successful breakout.[13]

Effects of Guidelines. When the NIH issued these guidelines in June 1976, they were received with expected ambivalence. Some scientists felt that they were prohibitively stringent; others insisted they were still too lax. Though the guidelines apparently meant that some experiments would be halted—such as those using animal bacilli as hosts—the NIH made it clear that it did not intend to enforce the guidelines with any special legal measures or penalties. The guidelines merely indicated that research proposals submitted for funding to the NIH would be jointly classified according to the dual designation system to indicate the level of safety procedures required. Thus, research came to be known among scientists as "P3-EK2" studies, or the like.

One immediate result of the new guidelines was the recognition by many universities and research centers around the country that they needed to construct new research laboratories. For recombinant experiments, P3 or P4 laboratories would be needed to satisfy the NIH grant reviewers, though only military installations or medical centers working with lethal pathogens would require the P4 rating. These laboratories were very expensive to build, and the difficulty of raising the necessary funds acted as a check on the widespread utilization of recombinant techniques in research. Nevertheless, many universities did initiate plans to build P3 laboratories. One such laboratory was proposed at Harvard University, and the prolonged public debate in the Cambridge, Massachusetts, city council that resulted from the proposal raised many new questions concerning research policy and the role of the NIH.

The Controversy in Cambridge

Much has been written about the dispute over the conduct of recombinant DNA research in Cambridge, Massachusetts. Not only did this public debate bring to a head the public's fear of a new scientific technique that sounded suspiciously like a real-life Frankenstein or Andromeda strain scenario, the debate also clarified the place of the NIH as a preemptive actor in securing the public safety. This controversy raised many issues concerning the policy regulation of potentially hazardous research, not all of which have yet been resolved.

The initial issue in Cambridge was the presumed right of scientists to freedom of inquiry unfettered by excessive governmental regulation. As Sheldon Krimsky elaborated,

> The federal government's role in regulating technologies and their byproducts is well established, but a clear distinction has been drawn between science and technology. Scientific inquiry is seldom justified on short-term utilitarian grounds, and freedom to pursue knowledge is highly valued. Thus, calls for regulation of science evoke extremely negative reactions both from scientists and from legislative bodies. The intervention of government into the affairs of the research scientist has been likened to a First Amendment violation—a restriction of the scientist's right of free inquiry.[14]

Of course, similar sentiments had been expressed by some scientists involved in recombinant research since before the Asilomar Conference, but the episode in Cambridge heightened fears because two new groups of actors had become involved: state and local agencies and an aroused lay public.

The controversy began early in 1976 when members of the biology department at Harvard proposed to turn several laboratory rooms into a P3 laboratory. Considerable resistance to the idea was mounted by other members of the department and was led by George Wald and his wife, fellow biologist Ruth Hubbard. The issue quickly spread across campus, and the university decided to hold a public meeting on the question in May 1976. Some members of the Cambridge City Council also attended the forum and subsequently urged the council to address the question as part of its obligation to protect the public safety.

The council held its first meeting on June 23—ironically the very day that the new NIH safety guidelines went into effect. It was attended by a huge and raucous crowd of scientists, students, and townspeople, and the adversary atmosphere was quickly established by the abrasive mayor, Alfred Vellucci, who was head of the council. Vellucci made little effort to hide his animosity toward the university in general and toward Harvard scientists in particular, whom he referred to as "those people in the white coats." Stating that "Cambridge has six square miles and we're the boss here," Vellucci castigated the university for not involving the city in its plans and concluded that "they're going to do what we tell them."[15] He closed the first meeting by proposing that all recombinant DNA experiments be banned within city limits for two years.

The Right of Free Inquiry. Notwithstanding the political grandstanding at the meetings, the council dealt seriously with several perplexing questions concerning scientific practices and policy making. The first issue, which had surrounded the research since it had initially been proposed in 1972, still required resolution according to many scientists and laypeople: the assumed right of scientists to freedom of inquiry in their professional pursuits. Where did that right come from, and how far did it extend in the face of possible safety risks to society? Other questions were also debated at the meetings, including whether scientists have special rights of inquiry not granted other citizens, and how and whether the goal of science granted that right. If the purpose of science was the pursuit of knowledge for its own sake, why were scientists given special latitude in the exercise of that right? If, on the other hand, the goal of science was to serve society, could not society legitimately restrict the exercise of the right of free inquiry if it deemed it in society's best interest to do so?

If indeed society had a right, and perhaps a duty, to regulate the rights of scientists to freedom of inquiry, how should that duty be exercised? The Cambridge meetings explored two aspects of this issue: The first focused on the role of public participation; the second on the question of federal preemption in policy making in light of the published NIH guidelines. First, should a nonscientific, lay public or its governmental representatives assume the authority to pass judgment on the highly sophisticated practices of normal science? The question of public participation was particularly troubling at Cambridge because it was the first controversy since the self-regulation of DNA research had begun at Asilomar and been carried through in the development of the NIH guidelines. How, it was asked, could a relatively uninformed public make intelligent decisions about a technology that it could only partially understand?

The opponents of the Harvard research laboratory had two answers to this question. First, they argued that since scientists themselves could not agree on the relative dangers of recombinant research, the decision must come from citizens and their representatives. Viewing the history of the development of the NIH guidelines, the mayor expressed through his aides the opinion that the disagreement among scientists before and after Asilomar disqualified the expert testimony by biologists on each side of the issue: "We looked at the process by which they arrived at those guidelines and found it was anything but placid. We were not reassured."[16] Furthermore, whether indeed scientists had the right to assume the responsibility of self-regulation—even if they could agree on the regulations—was not self-evident in the council's eyes, given the public stake in the outcome.

The second response of the council to the issue of lay involvement carried with it a new approach to the issue of who retains decision-making authority. Though citizens clearly do not have the expertise to make technical decisions, they would be affected by those decisions, and they argued that that circumstance in itself granted them a certain degree of authority. This basis for power in policy making is indeed entirely different from the two traditional arguments for policy-making authority derived from either professional expertise or political office. From siting disputes to biomedical technology, those who stand to use a new technology or be otherwise affected by it are claiming the right to pass judgment on it in public meetings, in referenda, and in citizen action groups.

Because of the demand for citizen input in the Cambridge dispute, the council claimed that its authority to protect public health outweighed that of federal preemption in policy making in this and similar matters. Though the NIH had published research guidelines, the historic judicial and congressional deference to state and local authorities in matters of public health granted the council, in its eyes, the authority to determine whether the experiments should take place within city limits. If the council's position were upheld, then the outcome in Cambridge would serve as a model for that in other cities faced with similar requests for P3 laboratory construction, a conclusion that the press emphasized and scientists ruefully acknowledged. Because they feared that the Cambridge example would indeed encourage other cities to set their own research regulations, many scientists during the controversy petitioned Congress to write legislation on the matter. Senator Edward Kennedy of Massachusetts did initiate legislation, as did other members of Congress, though as of 1986 none has been signed into law.

Challenge to the NIH. A second issue considered by the Cambridge City Council directly challenged the authority of the NIH to regulate recombinant research. After noting the disagreement among scientists within the NIH and those affected by its new guidelines, the council raised doubts about the efficacy of one federal agency serving both to promote (and fund) research in an area and at the same time to regulate it. Cambridge city councillor David E. Clem argued that the dual duties of the NIH were in fact contradictory and could not be performed by one agency in a manner sufficiently sensitive to the demands of protecting public health. He recalled the inefficiency of the old Atomic Energy Commission in attempting the same two functions and suggested that regulatory functions be transferred to state and local authorities.[17]

City Council's Decision. After much debate and testimony, the City Council met on July 7 to render its decision. In a council chamber overflowing with citizens and press, the mayor announced a three-month "good faith" moratorium on recombinant DNA experiments within city limits. Though he had originally pushed for a two-year ban, the town councillors had advised the council that three months was the limit of the city's legal authority. The mayor also announced the establishment of a special town committee to study the question of whether the research should be allowed after the moratorium ended. The Cambridge Experimentation Review Board (CERB) held hearings and drafted a report that it presented to another packed council meeting on January 14, 1977. To the chagrin of the mayor, the CERB "unanimously recommended to the city's acting health commissioner . . . that experiments with recombinant DNA be allowed."[18]

Though the recommendations of the CERB resolved the controversy in Cambridge, its actions seemed something of an anticlimax. Nevertheless, after the CERB report Cambridge scientists found that though they were free to conduct their experiments, they were also subject to a new set of guidelines established by the board. Thus the precedent was set for scientists: If they were to carry out their research, they would have to put up with considerable interference by the public and its local agents. However, the Cambridge example and its consequences were not emulated in many other U.S. cities or university towns. In April 1977, the Dane County Board of Supervisors for Madison, Wisconsin, passed a resolution in support of the NIH guidelines and thereby permitted DNA experimentation at the University of Wisconsin to be regulated solely by the NIH. Most other U.S. universities seeking permission from local officials to construct P3 laboratories found that those officials were more willing to follow the Madison example than that of Cambridge, in order to avoid the kind of controversy that had erupted in Massachusetts.[19]

After the prolonged battle in Cambridge, the controversy seemed to die away rapidly. Among the many reasons for this, three lend themselves to further analysis for their impact on the later development of the technology and policy regulating it: (1) the announcement in 1978 of the production of the first human hormone, somatostatin, using recombinant DNA techniques; (2) the Supreme Court's decision in the 1980 Chakrabarty case permitting the patenting of new life forms created in the laboratory using recombinant methods; and (3) the incredible rush toward commercial exploitation of the new technology, culminating in the 1980 public offering of stock in the new biomedical corporation, Genentech.

COURTS AND CORPORATIONS: DNA POLICY AFTER CAMBRIDGE

Since the initial discovery of techniques to recombine DNA molecules, those scientists favoring unimpeded development of the technology had promised discoveries that would revolutionize medicine and immeasurably advance medical treatment and disease prevention. By 1978, some of these promises had already been fulfilled. The production through recombinant techniques of the human hormone somatostatin signaled throughout the scientific world that the new technology was ready to bear fruit. Slightly more than a year later, the *New York Times* announced that laboratory scientists had created human interferon, a hormone believed useful in the prevention and treatment of cancer.[20] The announcement of the interferon experiments and subsequent positive publicity throughout the media had two consequences: They quelled almost instantly much of the public's lingering fear concerning recombinant technology, and they started a land rush business in venture capitalism to establish corporations to produce and sell the products of recombinant research.

By 1978 the NIH had relaxed its guidelines regarding recombinant experiments, and all congressional attempts to legislate additional regulations had failed because of pressure from both the NIH and the business community.[21] Thus the way was clear for commercial exploitation of the new technology without substantial governmental hindrance. Throughout the next two years numerous small corporations were established for research and development of the products of recombinant technology, and by 1982 corporate laboratories had successfully produced a variety of new products, including interferon, human growth hormone, synthetic insulin, and genetically altered bacteria, which actually fed on petroleum products, including oil spills. By November 1979, the paper value of the four largest of the new genetic corporations—Cetus, Biogen, Genex, and Genentech—was $225 million, and this value doubled in the subsequent six months.[22] When Genentech offered shares of stock for public purchase in September 1980, the price rose from $35.00 to $89.00 per share in minutes.

The surge of financial speculation and investment in recombinant genetic technology created a new industry, although by 1982 no products of the technology had yet reached the market. In other words, none of the new genetic technology firms had anything but paper value: Nothing had been sold that could provide profits and therefore a return on investment. Marketing the new products waited on several government decisions concerning their safety and their patentability. The first of these decisions had to come from the Food and Drug Administration; the second came from the Supreme Court on June 16, 1980, in the decision in the *Diamond* v. *Chakrabarty* case.

Diamond *v.* Chakrabarty. In the petition to the lower court of Customs and Patent Appeals, University of Illinois biologist Ananda Chakrabarty had sought a patent of a new bacterium created by recombinant techniques. The court upheld the application, but the commissioner of the U.S. Patents Office appealed to the Supreme Court to disallow the patent, citing a Supreme Court case in 1978, *Parker* v. *Flook,* that had encouraged the Patents Office to "proceed cautiously when asked to extend patent rights into areas wholly unforeseen by Congress."[23] In brief, the court decided that although the new bacteria were indeed alive, they were the products of human beings, not of nature, and were therefore patentable. Thus, it upheld the reasoning of the Court of Customs and Patent Appeals, that "in short, we think the fact that micro-organisms, as distinguished from chemical compounds, are alive, is a distinction without legal significance."[24]

Patent Rights. The decision in the Chakrabarty case removed one roadblock in the path of commercial exploitation of genetic technology, but other questions pertaining to patent rights remained unresolved by the Patents Office. These questions derived from two aspects of the technology and extended to virtually all areas of scientific research. First, almost all genetics research since 1950 was supported by public funds granted to either university or government laboratories. In light of this history, some scientists raised the issue of whether patents should be granted to private corporations or individuals for profit-making enterprises. Most early patents for recombinant techniques developed in university laboratories were issued to the universities themselves, and the techniques were freely available to anyone who wanted to use them for research. However, as corporations began to apply for patents in 1980, it was understood that they would retain exclusive rights to the technique, as they were competing with other firms for profits from the use of the techniques or from the sale of the patents. Thus, universities or scientists operating under public funds found they were at a disadvantage in pursuing research because certain patented techniques needed to further their work were unavailable.

Second, the granting of patents for recombinant techniques challenged certain conventions of the scientific community. The awarding of a patent presumes that the grantee invented the technology; however, inventorship is difficult to isolate in a joint research effort in which scientists at several university or outside laboratories build on the research results of others. As Stanford biologist Stanley Cohen pointed out, "Scientific advances such as the one we have been involved in are in fact the result of multiple discoveries carried out by many individuals over a long period of time."[25] Within the scientific community, inventorship is conventionally indicated when a scientist publishes a research report; however, when such articles are co-authored by any number of scientists at different institutions the legal definition of inventorship is difficult to determine. Although co-authorship frequently has been a matter of courtesy in the scientific world, it now may mean that the co-author becomes a holder of a very valuable and financially rewarding patent. Thus the financial exigencies may mean a fundamental change in the manner in which scientific research is conducted and in who is given credit for that research.

Corporate Secrecy. The economic exploitation of recombinant research has affected the scientific community in other ways. Prominent scientists in genetics have been wooed away from academia into corporate laboratories since the commercialization of gene recombination first began, and some scientists fear what this will mean for the free exchange of knowledge that characterizes science in an academic setting. One such fear would directly affect policy making: Corporate scientists may become less willing or less free to give advice to local policy makers concerned with the conduct of research, as the academic scientists did in Cambridge. Doing so might jeopardize corporate secrets and cause the loss of a competitive edge in the rush to exploit the new technology. A second fear within the scientific community would be that the increasing commercialization of recombinant technology will "allow commercial interests to influence the goals and nature of academic research." Sheldon Krimsky of Tufts and the RAC commented: "Just as war-related academic research compromised a generation of scientists, we must anticipate a similar demise in scientific integrity when corporate funds have an undue influence over academic research."[26]

Krimsky might have been overstating his case, but his fear is reminiscent of that expressed by scientists in the Cambridge controversy. In Cambridge, scientists worried that public scrutiny of their research in the attempt to regulate it would violate their constitutional right of freedom of inquiry. Krimsky feared the same result in the commercialization of science. If scientists could no longer present their research results at conferences or in published articles for fear of giving away company secrets, the free exchange of knowledge— an ideal of science—would be in serious jeopardy. Although the attempt to regulate locally recombinant research in Cambridge did not result in the limitation of scientific inquiry by laypersons acting as concerned citizens, such limitation may indeed result when laypersons act as the heads of new genetic-technology corporations.

The new aura of secrecy surrounding recombinant experiments has recently been extended to the policy process regulating the research. As stated before, the only guidelines relevant to the research are those provided by the NIH, and the agency's regulatory power extends only to experiments conducted under NIH funding. This arrangement has put much university research at a disadvantage because the NIH provides much of the money to conduct the experiments at university laboratories. Although experiments by corporations and others using non-NIH funds are exempt from formal government regulation, in many cases the scientists conducting the research have voluntarily submitted their proposals to the NIH for approval. That approach may be changing, however, as a direct result of the need for corporate secrecy in developing the technology.

Challenge to the RAC. In October 1983, two industry proposals submitted for NIH approval were accompanied by a request that the meetings to consider the proposals be held behind closed doors. The NIH complied with the request for secrecy on the grounds that proprietary data were included in the request and in the discussions of the research. However, the secrecy caused a stir in the industry and among university scientists and raised the issue of how far RAC regulatory power extended.[27] This question was partially answered later in 1983 when federal court judge John J. Sirica ordered that recombinant research approved by the RAC and funded by the NIH be halted on the grounds that it involved a release into the atmosphere of bacteria with recombined genetic material. However, the ruling specifically exempted experiments conducted by private companies using nongovernment funding.[28]

RAC authority was challenged even further when the Monsanto Company requested clearance from the Environmental Protection Agency (EPA) for similar experiments and completely ignored the RAC for fear that its research plan would be published. The traditional reliance on NIH guidance—if not actual policy—in the field of genetic recombination research appears to have eroded to the point that only research at universities under NIH aegis is subject to federal scrutiny of any kind. All other research and experimentation are utterly free of regulatory policy and are conducted largely in secret. Likewise, universities are relying more and more for financial support on private corporations interested in the eventual profits from research applications. Thus the free discovery and exchange of scientific knowledge— even in the hallowed halls of academe—are being challenged on two fronts: Scientific research in some instances is being conducted (1) for profit and (2) in secret.

CONCLUSION

The commercial exploitation of genetic recombination technology has proved to be both a blessing and a challenge to the community of science. By itself, this transfer of a new technology from the rarified world of university laboratories operating under public funding to the boardrooms of private corporations is a positive example of the movement of knowledge in a free society. Such transfers have a long history in the United States, and through the sharing and exploitation of new knowledge science is made to work for the betterment of society. Furthermore, the entry of corporations into the field of genetic research has meant more research dollars flowing into laboratories in both private and public sectors, and this movement has accelerated the gaining of knowledge about the genetic makeup of life. Thus, the alternative goal of science as "knowledge for knowledge's sake" has also been served by the commercial history of gene recombination.

The challenges presented to science by public interest in the manipulation of genetic material have been numerous, but most have focused on balancing regulation of genetic research to protect the public with assurance of freedom of inquiry for those engaged in the research. This balance has not been easy to achieve. As seen in Cambridge and other towns in which construction of P3 laboratories was proposed, the public feared the possible dangers of recombinant research, whereas scientists worried that public scrutiny would mean the demise of their experimental research. These conflicting anxieties appear to have largely disappeared, however, and the technology has rapidly developed and expanded.

The history of public involvement in the case of recombinant DNA indicates that though the public and its political representatives are frequently taken aback by the ever growing impacts of science on society, they also understand and appreciate the benefits that science offers. The scientists' fear of public interference in the DNA controversy proved to be unfounded; as Sheldon Krimsky concluded, the effects on science of stringent public scrutiny were negligible in terms of the restriction of scientific research.

> The controversy over the use of recombinant DNA molecule research highlights many problems in the relations between science and society. One of the most crucial issues, however, revolves around the belief in the necessity of scientific autonomy. . . . But nothing much did, in fact, happen. Citizen groups did not ban the research and Congress has not strangled molecular biology. Church organizations did not renounce genetic manipulation as a violation of God's will. Books were not burned. On the contrary, some citizens began reading more about science than they had before.[29]

A more persistent challenge to scientific autonomy was presented by the entry of corporations into the field of science, but here too the development of recombinant technology demonstrates the strength of this scientific ideal. Though considerable secrecy exists in the corporate development of genetic techniques, the patenting process itself ensures eventual publication of new scientific advances. In addition, the demands of capitalistic success have considerably accelerated the exploration of the secrets of DNA. On both counts then, science has not noticeably suffered from the public and corporate presence in its community. If a threat does exist to science's right of free inquiry, it perhaps comes as much from the scientists' eagerness to allow commercial exploitation—and therefore from the degree of secrecy that competition entails—as from the watchful eyes of citizens.

As a final point of interest, the development of genetic technology is notable for the distinctiveness of the policy making it generated. It is not unique because little federal policy has been established relevant to the technology—that holds true for other technologies discussed in this book. The distinctive feature in the development of policy making for genetic technology is the combination of self-regulation and local regulation.

At the Asilomer Conference the scientific community showed an awareness of risk and a self-restraint that were distinctive and laudable manifestations of its self-regulation in the public interest. The actions of the Cambridge City Council and the consequences of the prolonged public dispute there manifested the same beneficial awareness. It is provocative and perhaps instructive to consider whether the policy-making mechanisms at these two events constitute a single, innovative policy process for science and technology decisions. These two examples show both scientists and citizens taking active parts in generating policy for a new technology. In each instance, both groups stood to benefit and to be challenged by the new technology; thus, both took active roles in the policy process: scientists as practitioners and teachers, citizens as potential users and willing students. Though each group was initially wary of the other, the product of their joint involvement in making policy served to reassure and benefit both. Working together they joined the interests of science and society, and their combined efforts made up a successful process for making technical decisions. The process may have been tempestuous and at times raucous, but these characteristics often arise in a democracy when interests and convictions run deep, and it can be argued that in a democracy, this is how it should be.

NOTES

1. *Science* 182, no. 566 (November 9, 1973).

2. *Science* 185, no. 303 (July 26, 1974).

3. See, for instance, *New York Times,* July 18, 1974; and *Washington Post,* July 18, 1974.

4. James D. Watson and John Tooze, *The DNA Story* (San Francisco: W. H. Freeman, 1981), p. 25.

5. *Ibid.,* p. 26.

6. See, for instance, Michael Rogers, *Biohazard* (New York: Knopf, 1977).

7. Watson and Tooze, *op. cit.*

8. *Science* 185, no. 303 (July 26, 1974).

9. Watson and Tooze, *op. cit.,* pp. 63–64.

10. *Ibid.,* p. 64.

11. *Ibid.,* pp. 65–66.

12. The following summaries are taken from "Recombinant DNA: NIH Group Stirs Storm by Drafting Laxer Rules," *Science* 190, no. 767 (November 21, 1975), pp. 767–768.

13. *Ibid.*

14. Sheldon Krimsky, "Regulating Recombinant DNA Research," in Dorothy Nelkin, ed., *Controversy: The Politics of Technical Decisions* (Beverly Hills, Calif.: Sage, 1979), p. 277.

15. *Washington Post,* July 9, 1976.

16. "Recombinant DNA: Cambridge City Council Votes Moratorium," *Science* 193, no. 300 (July 23, 1976).

17. *Ibid.*

18. *The Real Paper* (Boston), January 15, 1977, p. 1.

19. "Gene-Splicing: At Grass-Roots Level a Hundred Flowers Bloom," *Science* 195, no. 558 (February 11, 1977).

20. *New York Times,* January 17, 1980.

21. Watson and Tooze, *op. cit.,* chaps. 6 and 12.

22. *Ibid.,* p. 488; *Science* 208, no. 688 (May 16, 1980).

23. *Parker* v. *Flook,* 437 U.S. 584 (1978).

24. *Ibid.*

25. "Inventorship Dispute Stalls DNA Patent Application," *Nature* 284, no. 388 (April 3, 1980).

26. *Ibid.*

27. "NIH to Review Policy of DNA Committee," *Science* 223, no. 35 (January 6, 1984).

28. "Rifkin Broadens Challenge in Biotech," *Science* 225, no. 297 (July 20, 1984).

29. Krimsky, *op. cit.,* p. 249.

SELECTED READINGS

Beers, R. F., Jr., and E. G. Bassett, eds. *Recombinant Molecules: Impact on Science and Society.* New York: Raven Press, 1977.

Goodfield, J. G. *Playing God: Genetic Engineering and the Manipulation of Life.* New York: Random House, 1977.

Grobstein, C. *A Double Image of the Double Helix: The Recombinant-DNA Debate.* San Francisco: W. H. Freeman, 1979.

Hutton, R. *Bio-Revolution: DNA and the Ethics of Man-Made Life.* New York: New American Library, 1978.

Jackson, D., and S. Stich. *The Recombinant DNA Debate.* Englewood Cliffs, N.J.: Prentice-Hall, 1979.

Judson, H. F. *The Eighth Day of Creation.* New York: Simon and Schuster, 1979.

Richards, J., ed. *Recombinant DNA: Science, Ethics and Politics.* New York: Academic Press, 1978.

Rogers, M. *Biohazard.* New York: Avon, 1979.

Wade, Nicholas. *The Ultimate Experiment.* New York: Walker and Company, 1979.

Watson, J. D. *The Double Helix.* New York: Atheneum, 1968.

Watson, James D., and John Tooze, eds. *The DNA Story.* San Francisco: W. H. Freeman, 1981.

8

BIOMEDICAL TECHNOLOGY: SCIENCE, FREEDOM, AND PERSONAL MORALITY

In the previous chapters we have seen that when new scientific knowledge is used in the creation of new technologies, new choices and dilemmas of policy making quickly follow. However, sometimes the mere existence of new knowledge, regardless of how or even whether it is used, has the same effect. For instance, the possession of the knowledge of how to construct nuclear weapons is independent of the actual manufacture or use of those weapons. Even if all countries that possess nuclear arsenals destroyed them tomorrow, the realization that the knowledge existed and could be used would continue to affect both international relations and the strategies of war.

In this chapter we will examine the impacts of new knowledge—impacts that create dilemmas of moral choice felt at both the personal and policy level. Biomedical technologies that supply new information concerning the genetic heritage and health risks of adults and fetuses offer applications that are intrinsically controversial on moral grounds for both individuals and policy makers. As with knowledge about nuclear weapons, the knowledge these technologies provide raises the same controversies whether or not doctors or patients choose to apply it. The knowledge is itself at issue, as well as its potential uses. Consider the following case in Melbourne, Australia.

Elsa and Mario Rios, a young and very wealthy couple from Los Angeles, were devastated by the death of their ten-year-old daughter in 1978. In 1981 the government of Australia invited them to participate in a government-financed fertility program in Melbourne involving artificial insemination and cryopreservation (freezing of the fertilized egg, or zygote). Having searched in vain for a way to have another child, the Rioses eagerly accepted and were the last foreign nationals admitted into the program. At the Queen Victoria Medical Clinic in Melbourne, physicians extracted three of Elsa Rios's eggs and fertilized them in a laboratory dish using sperm from Mario Rios. The doctors then implanted one of the eggs in Elsa's womb. The procedure resulted in a miscarriage ten days later. The other two fertilized eggs were frozen in liquid nitrogen and preserved for implantation at a later date. In April 1983 the Rioses were killed in a plane crash in Chile. They left no will and no heirs.

Whether the frozen embryos constituted heirs to the substantial fortune left by the Rioses was the question London newspapers posed after the couple's death. Though the issue seems preposterous and certainly bizarre, the fate of the frozen embryos became an international cause célèbre as newspapers around the world picked up the story. And in Sidney, Australia, women lined up to have the embryos implanted in their wombs so that they could become adoptive mothers.

The Rios case was complicated by the fact that the frozen zygotes were almost certainly not viable, according to most experts, as the technique for cryopreservation was relatively primitive in 1981. Still, the disposition of the fertilized eggs was a question in a legal vacuum under both Australian and U.S. law. Because the Rioses were U.S. citizens, their lawyer, Laura Horwitch, asked the Los Angeles Superior Court to determine the legal status of the embryos. The case has not yet been finally decided and indeed may never reach resolution until legislatures write statutes that deal with the host of issues raised by this bewitching case.

Cryopreservation of fertilized eggs is now being practiced around the world in approximately two hundred fertility clinics. The eggs are always fertilized in vitro (outside of the womb in a lab dish) and result in pregnancy after implantation in approximately 20 percent of the cases.[1] In the United States, neither Congress nor state legislatures have passed laws regarding in vitro fertilization, but guidelines have been drawn up by the American Fertility Society to deal with cases like that of the Rioses. Under these guidelines, doctors must not sustain cryopreservation of fertilized eggs after the death of the genetic mother, or after she has become too old to give birth naturally, and in any case, the doctors and patients must jointly decide the disposition of the frozen eggs prior to the procedure. The eggs can be destroyed, used for research for up to fourteen days, or offered for adoption, in which case absolute anonymity is mandatory.[2]

Although such guidelines would seem to resolve a controversial case if it occurred in the United States, many factors might intervene to block a decision. Though guidelines exist, they have neither the sanction of law nor of society, since they have never been legislated or tested in court. Furthermore, though the Department of Health and Human Services in 1979 recommended that in vitro fertilization and cryopreservation be considered "ethically defensible" procedures, the politics surrounding reproduction and abortion make this recommendation weak and hardly suitable as public policy.[3] Finally, these procedures are only a part of a growing group of new technologies in the field of human reproduction and the manipulation of fetuses prior to birth. Almost all these technologies are currently available, and U.S. law regulates almost none of them.

BIOMEDICAL TECHNOLOGY AND INDIVIDUAL CHOICE

Biomedical technology is a generic term denoting an array of procedures and equipment in use or contemplated for use in the near future. Generally, this term covers everything from surgical techniques like artificial limb and organ replacement to cloning. In this chapter we will discuss two controversial technologies, often employed in conjunction with each other, that have widespread legal and policy implications. The first technology is that of genetic screening and its uses in the discovery of inheritable diseases. The second is the whole area of reproductive technologies, including artificial insemination, sperm banking, amniocentesis, ultrasound scanning of fetuses, in vitro fertilization, and cryopreservation of embryos. We will begin with a brief discussion of each technology and its uses in disease control and identification and in human reproduction.

Genetic Screening

Genetic screening was developed from extensive research on the recombining, or splicing, of human genes. As explored in Chapter 7, the research created much controversy in Cambridge, Massachusetts, and Bloomington, Indiana, where recombinant DNA work was begun in the early 1970s. Though the fear of releasing mutant Andromeda strains through recombinant research has died down, the worries accompanying the uses of this knowledge of the human gene have not been totally laid to rest.

Screening, which detects the presence of mutant genes that result in disease or the transmission of disease to offspring, can be performed at any stage of life from before birth to death. In the prenatal stage the process requires amniocentesis, which will be discussed in the next section. Postnatal screening usually involves the microscopic examination of chromosomes in blood cells removed during routine blood sampling. Information from screening provides insight into a vast array of diseases and malformations that have affected individuals since the origin of the species, as well as some that have appeared relatively recently.

Currently, between one hundred thousand and two hundred thousand infants are born annually in the United States with congenital malformations, single-gene hereditary disorders, or chromosomal anomalies—the three general categories of genetic disorders tested for in genetic screening. The number of occurrences of these disorders represents approximately 3 to 5 percent of the 3 million live births in the United States, and together they are responsible for over 20 percent of all infant deaths.[4] Infant diseases and malformations constitute a large portion of what is called genetic disease, though diseases in that category affect adults as well. Infants may be afflicted with genetic disorders such as phenylketonuria (PKU), Down's syndrome, and spina bifida, or by genetic diseases like sickle-cell anemia, Tay-Sachs disease, heart disease, and Huntington's chorea. Indeed, many scientists and doctors believe that if genetic screening were employed as a routine medical test, it might uncover the causes of most debilitating human diseases and disorders.

Perhaps the most controversial use of genetic screening techniques is to detect carriers of disease. Many diseases such as sickle-cell anemia or Tay-Sachs can be carried and transmitted by individuals who do not suffer from the disease. Many doctors regard carrier status as important information, though the course of action based on that knowledge is difficult to decide upon. The existence of the knowledge itself raises dilemmas about who should have access to carrier information and what the consequences of such access would be. The law and even traditional morality have little to say on these questions, but as we shall see, the answers have critical impact on the individual's life and on future generations.

Artificial Insemination and In Vitro Fertilization

Artificial insemination (AI) and in vitro fertilization are distinct reproductive technologies, which doctors frequently use together to enable couples to conceive when either the prospective mother or father has an inadequate reproductive capacity. Artificial insemination is not a new technique; farmers have used it for centuries in animal husbandry. Even among humans artificial insemination dates back to the eighteenth century, though its use has been widespread only since the 1940s.[5] Today artificial insemination is the most widely employed human reproductive technology, and some researchers have estimated that from ten thousand to twenty thousand Americans employ the technique annually.[6]

In vitro fertilization, commonly known as test-tube conception, involves extracting an egg from a woman, fertilizing it by male sperm, and subsequently replacing the fertilized egg (zygote) in the mother's womb or that of a surrogate mother. This technique has raised more legal issues than other techniques because it sometimes pits a genetic mother against a biological mother (who provides the womb), and a genetic father against a social father. Because this technique is always used with artificial insemination and frequently in conjunction with other reproductive technologies such as cryopreservation or sperm banking, or surrogate mothers, it is fraught with all the questions surrounding those technologies as well. Since in vitro fertilization was first successfully employed in 1978, the technique has been improved greatly in terms of numbers of successful implants and of children brought to term. Today it is used more often and with considerably less public fanfare than five years ago, and it has been a great boon to women with reproductive tract abnormalities and to men with low sperm counts.

Together artificial insemination and in vitro fertilization constitute a great breakthrough in reproductive capability for many people who once could not conceive children. However, they carry with them new demands—which vary according to a country's cultural traditions regarding reproduction—that cannot easily be addressed within the ethical and legal framework of a culture.

Fetal Diagnosis and Sex Preselection Technologies

Of all the new reproductive technologies, one of the most controversial is that used in the prenatal diagnosis of fetuses. Members of Congress have referred to these technologies as "search and destroy missions," and the techniques are haunted by what L. R. Kass called the "ghost called the morality of abortion."[7] The most well-known of these technologies is amniocentesis, a procedure in which a doctor withdraws approximately 20 cubic centimeters of fluid from the amniotic sac surrounding the fetus and examines the cells in it for genetic malformities or disease. Amniocentesis is a highly informative prenatal test and can detect "virtually all chromosomal abnormalities in the fetus, approximately 75 serious inborn metabolic disorders, and approximately 90 percent of fetal neural tube defects."[8]

The medical community, supported by the statistical results of scientific tests, have declared the technique to be safe for both mother and child. However, the furor over the use of amniocentesis continues because the technique and other diagnostic technologies like ultrasound scanning of the fetus logically presume that abortion is an option pending the test results. Though a woman might conceivably want to know the facts about her unborn child even if abortion were not an option, the test would probably be used infrequently for such cases.

The prospect of abortion as a consequence of amniocentesis has frequently been overemphasized by critics of abortion. According to a study by the National Institutes of Health in 1979, the argument that amniocentesis leads automatically to abortion is simply not borne out by two facts. First, only 3 to 5 percent of all amniocenteses result in abortion. Second, the evidence from obstetricians suggests that many women at risk in having children would either automatically have an abortion in the case of pregnancy or refrain from becoming pregnant at all if prenatal diagnostic techniques were not available. In fact, more couples are continuing pregnancies because of encouraging results of amniocentesis, even though they know they are at risk, than are aborting fetuses because of the results of the test.[9]

Other prenatal diagnostic tests include ultrasound scanning, fetoscopy, placental aspiration, and fluorescent staining. These techniques are important for the information they provide concerning the health of the fetus, especially if they indicate the need for fetal surgery—another recently developed technique. But like amniocentesis they all provide one additional bit of information that is potentially powerful in its impact: the sex of the fetus. All these techniques are potential modes of sex selection and as such are controversial. Although most doctors would not perform the tests simply to determine the sex of the fetus, that information is forthcoming from all the tests and thus must be provided upon the parents' request.

The implications of sex selection techniques are obvious, although surveys conducted in the United States suggest that adults do not prefer children of one sex over another.[10] In other countries, the bias in favor of male offspring is still very strong; thus the techniques may be abused and only female fetuses aborted. Aborting a fetus simply because it is not of the desired sex may seem drastic, and indeed it is both frivolous and dangerous given the risks of abortion at any stage of pregnancy.

Sex preselection techniques, which circumvent the need for such an abortion, are less well known and less often used than amniocentesis, but widespread awareness of their existence and effectiveness seems inevitable. Two of these techniques are currently being tested, neither of which involves the issue of abortion because they are employed before the start of the reproductive process. The first employs a centrifuge, in which male semen is placed and spun at high speeds. The whirling action separates the X sperm from the Y sperm, the first of which carries the female trait and the second the male. After the sperm has been segregated, an egg is fertilized with the sperm carrying the desired trait (either through artificial insemination or in vitro fertilization).

The second technique for sex preselection is even more passive in the sense that it requires less sophisticated intervention or fewer trips to the laboratory. A doctor places a special diaphragm of a superfine mesh material in the prospective mother's vagina. The fineness of the mesh allows only Y sperm to pass through it; X sperm are blocked because they are larger than Y sperm. The technique obviously requires neither artificial insemination or in vitro fertilization though it can be used in conjunction with artificial insemination. Equally obvious is the fact that this technique can only be used for the conception of a male fetus.

Of all the techniques for sex selection and preselection, the last is the easiest to use and the least intrusive. It is also the technique that best guarantees male offspring. As a result, it has proved to be troublesome to doctors, researchers, and feminists. The widespread use of this technology might cause serious social changes in societies in which male offspring are particularly desired.

BIOMEDICAL TECHNOLOGY AND MORAL CHOICE

Even the preceding brief description of new genetic and reproductive technologies raises numerous questions about the morality of new biomedical techniques. These questions go to the heart of current political and ethical belief systems and are perhaps by nature unanswerable in the sense that disagreement over their implications will probably never end. However, discussion of the ethical dilemmas raised by biomedical techniques is important because through it we can attempt to make viable individual or policy decisions.

Each of the genetic and reproductive technologies raises general issues of rights: who has them, to what extent society or political authority should protect them, what happens when individual rights conflict with each other or with those of others, and what are the rights of society as a whole, of the unborn, and of future generations? Although some of these issues are questions of public policy, all are quandaries of personal moral choice.

Individual and Social Aspects of Screening

Genetic screening raises issues of individual and social rights in several contexts. Individuals have the right to make decisions privately concerning their own health whereas society as a whole has the right to prevent the spread of disease. Screening, as Blank pointed out, has a distinctly individualistic orientation: It focuses on identifying and reducing the incidence of genetic disease and in some cases on suggesting therapy.[11] However, genetic screening also has a social dimension: It is used to prevent disease or to lessen the spread of certain diseases—the goals of all public health programs. Sometimes this social concern is widened to include the welfare of the species as a whole, and genetic screening programs are advocated to strengthen the genetic base of a society or of the race.

Collective motivations are different than the desire to help individuals; yet the technology is the same, as in some sense are the reasons to employ it. However, the clash of the assumed individual right to reproduce with the more nebulous social obligation to protect society's gene pool is often unavoidable. For instance, during a routine premarital blood test a doctor might discover that a prospective mother or father is a carrier of Tay-Sachs disease or sickle-cell anemia. Does society have the right to prevent the couple's conceiving a child—and therefore prevent their passing on either the trait or the disease? Or does the couple have a right to have children if it chooses? Such cases raise the spectres of eugenics as well as of enforced sterilization, and it is certainly unclear whether society has the right (or obligation) to engage in these practices no matter how laudable its concern for public health. Is the right to reproduce an inalienable right of an individual, or was Justice Holmes correct in his harsh dictum in an early sterilization case involving a mentally retarded woman that "four generations of idiots is enough"? Policy is understandably difficult to formulate in this area of public health, particularly within a democratic political system: It must incorporate the viewpoints of both individual conscience and societal obligation.

Genetic screening for inheritable disorders raises the even murkier question of the obligations present societies must shoulder to protect the rights of future generations. In discussions of energy many argue that those living today do not have the right either to deplete fossil fuel resources or to pollute the environment with nuclear wastes. Although these energy-policy quandaries may seem difficult, they pale in comparison to balancing the reproductive rights of the living with those of the not yet born. This issue goes beyond the question of abortion: It includes not only the rights of the unborn to live but also their future rights to reproduce without fear of passing on debilitating genetic diseases. Do future generations have the right to impose such obligations on the living, obligations that by their nature would abrogate the rights of some individuals to reproduce because of their carrier status? These obligations cannot be borne by all living persons of reproductive age, since not all are carriers. Thus, not all members of society would have to sacrifice to protect future generations (this is also more or less the case in the energy example). The prohibition forbidding some to reproduce involves not only a clash of rights—though it is a prodigious one—but also a question of equity, justice, and discrimination.

The rights of future generations and our obligation to respect them indicate the paradoxical responsibility that sometimes comes with knowledge. If we have the ability through carrier screening to prevent the transference of genetic disease, are we obligated to do so? Is an individual notified of his or her carrier status acting immorally in conceiving a child? (This latter question has led to some curious legal disputes such as "wrongful birth" and "wrongful life" suits, which will be discussed in this chapter.) Are these questions even legitimately points of law? Does not the right of privacy, albeit a "penumbral" doctrine in Justice Holmes's words, protect individuals when such weighty decisions have to be made? And does not the right of privacy ensure that the individual alone has the right to make the decision regarding reproduction? The *Roe* v. *Wade* case which made abortions legal, seems to support the right of privacy in questions of reproduction; however, the issue is the extent to which it does so. At issue then are not only what decision should be made concerning genetic screening and the rights of future generations but also who should make it.

Scope of the Right to Know

Besides giving rise to the mind-boggling issues of individual rights versus societal obligations, screening techniques have been used increasingly by both public and private institutions to gather data about individuals, primarily (one assumes) for employment information. Some businesses and corporations, going beyond tests for ability or managerial qualities, have without scientific justification screened for genetic disorders that allegedly might affect performance. Similarly, the Air Force has disqualified applicants for flight training because genetic screening has indicated that they have sickle-cell anemia, a disease that some researchers believe is correlated with fainting at high altitudes or under stress. Resulting civil suits have argued that such testing invades the right of privacy of the applicants, and disqualification on these grounds counts as unlawful discrimination. The discrimination charge is also made because genetic diseases frequently affect specific and exclusive groups of people; for example, sickle-cell anemia affects only blacks. Because virtually every ethnic and racial group can claim its own special disease,[12] knowledge of human genetics could lead to new forms of discrimination, especially since answers to any questions involving conflicting rights will involve the lessening of one or the other side's rights.

Finally, genetic screening raises an issue rarely considered before: the right of individuals not to know. Similar in some respects to the right of privacy, this right is especially relevant in the case of prenatal screening for birth defects. Do parents have a right not to be told that their child will be born with genetic defects? In 1975 the National Academy of Sciences mused whether "compulsory screening of prospective parents before conception or directly of the fetus prenatally *forces* information on the parents that they may take into account in their childbearing decisions" (emphasis added).[13] Even if the decision to abort is left to the parents, is the granting of knowledge of potential birth defects a coercive or prejudicial intrusion into a private ethical decision for the parents? Or in any moral dilemma, is more knowledge always good?

The right not to know the results of nonvoluntary genetic screening tests carries over to other issues raised by the reproductive technologies described earlier in this chapter. Again, the clash of individual rights and the rights of society is apparent. Amniocentesis may provide knowledge of birth defects in a fetus that, if brought to term, would require not only love and understanding from parents but considerable expense by society as a whole. Medical resources must be made available to the child, and the parents may not be able to afford them. In some cases hospitals supply the services free; in others the bill is paid by insurance companies or by the government. However the expenses are met, all citizens share the costs, through higher hospital bills, insurance premiums, or taxes.

Obviously, the awareness of the presence of birth defects in a fetus while still in utero raises an ethical dilemma for some woman carrying such a fetus. Such a moral choice is itself one risk of new knowledge for any prospective parent. With the legalization of abortion, the presumption of privacy in this decision is clear, but the interests of society as a whole are also present, and therefore techniques regarding reproductive choice are policy issues as well.

Changing Traditions

The impact of all the reproductive technologies such as in vitro fertilization, amniocentesis, and ultrasound scanning extends beyond the now enlarged area of individual choice. This extended area is the realm of public policy, which as yet is rather barren of policies that define the limits of individual choice and of social concern. We will examine this lack of policy for reproductive technologies in the next section. Before turning to this discussion, however, one additional effect of reproductive technologies must be examined. This effect is so potentially disruptive to traditional cultural values and definitions that social decision is essential; yet it is so perplexing that any policy formulation seems unlikely.

When taken together, the genetic and reproductive technologies discussed in this chapter represent more than just opportunities for reproductive freedom for individuals. They pose serious problems in the definition of traditional concepts such as motherhood, fatherhood, and even humanhood. For instance, by using in vitro fertilization and subsequent implantation of a zygote in a woman's uterus, the definition of fatherhood or paternity becomes split into two: One may be either a genetic father or a social father, depending upon whether the egg was fertilized by the woman's spouse who will help rear the child, or by the sperm of a donor. Similarly, if the egg is not supplied by the woman in whose uterus it is placed, the woman can be said to be the biological mother and the social mother (if she then adopts and rears the child) of the subsequent baby, but not the genetic mother, because the egg was supplied by someone else. The genetic mother is still the mother in some sense, although not in any sense usually acknowledged by cultural mores. Similarly, a surrogate can be used to carry the fetus: The fertilized egg of genetic parents can be implanted into the uterus of another woman who will carry the child to term, and in so doing can perhaps claim those rights concerning the child that issue from being the biological mother.

These complications are bewildering precisely because they alter the common-sensical and traditional concepts of motherhood and fatherhood held by any society. Some cases involving surrogates have led to protracted legal disputes over who indeed has rights of maternity. However, even more perplexing possibilities may arise. The use of surrogates seems to some feminists and others to raise the possibility of the creation of an entirely new class of women defined by their role—or occupation—of leasing their bodies for nine-month stretches. Some feminists view this approach as demeaning and suggest that it could lead to the degradation of women as merely reproductive machines. At the same time they understand that surrogacy is a logical extension of a woman's guaranteed power over her own body, as argued by proabortionists.[14]

From another perspective, reproductive technologies offer women possibilities for self-fulfillment never before contemplated. Though the medical community has viewed artificial insemination since its first human application as a corrective for male impotence, the technique can also be seen as a method by which a woman may have a child without involvement with any particular man. Similarly, by using a surrogate, a woman might have a child without ever having a relationship with a man or going through pregnancy. Feminist Barbara Rothman contemplated the cultural impacts of such possibilities with an ambivalence that marks the depth to which reproductive technologies can alter thousands of years of human sexual history and the cultural attitudes about reproduction. In an article written for *Ms.* magazine, Rothman applauded the new technologies for providing women additional rights that she calls "the rights of paternity"—the same rights as men have concerning reproduction—mainly the right not to be forced to carry the child to term.[15] However, in a later book Rothman suggested that reproductive technologies benefit women only if they are willing to deny their nature as women—as beings uniquely able to fulfill their nature by the act of carrying a child and giving birth. In this sense, the "rights of paternity" are merely a trap that forces women to become like men in order to achieve equal rights.[16]

Rothman's ambivalence about surrogacy and artificial insemination is indicative of the momentous changes in cultural attitudes regarding human reproduction brought about in part by new biomedical technologies. It also shows us how difficult it has been and will be to make policy in these areas. Because reproductive technologies are being introduced at a rapid pace policy makers cannot keep up. Nor can private citizens make up their minds about the advisability (moral, cultural, or otherwise) of employing genetic screening, artificial insemination, surrogacy, or in vitro fertilization. Amid this uncertainty, policy decisions must be made, but it remains to be determined how and by whom.

BIOMEDICAL TECHNOLOGY AND PUBLIC POLICY

The issues surrounding biomedical technologies do not easily lend themselves to political or policy resolution: Artificial insemination, surrogacy, and genetic screening are laden with ethical and ideological baggage that discourages policy makers from confronting them. Policy makers in legislatures, government agencies, or the courts are ill at ease when asked to resolve issues of individual freedom, moral choice, and individual rights, particularly when some of the rights belong to the unborn. In fact, an observer of biomedical policy argued that the agents of the U.S. political system are fundamentally incapable of making decisions concerning biomedical technology and are likely to remain so.[17]

State and Federal Policy

Before agreeing with this discouraging conclusion, we should examine the efforts that have thus far been made within the U.S. policy-making system. Efforts in the form of legislation or agency regulation are few, although much study has been done by national agencies on the impacts of biomedical technology. In the late 1970s the Office of Technology Assessment published two reports that acknowledged the potential ethical issues arising from genetic screening and in vitro fertilization. Although Congress has not passed legislation on these matters, in 1978 it held hearings on the impacts of biomedical technology in the Senate Committee on Human Resources and in the House Committee on Science and Technology.[18] Unfortunately, as Blank pointed out, these hearings suffered from the usual maladies of committee meetings: poor attendance, unbalanced testimony, unequal access for all viewpoints, limited distribution of the findings, and the lack of real debate. No reports or recommendations were forthcoming.[19]

Significant policy investigation and formulation have simply not taken place either in Congress or in executive agencies through the first term of the Reagan administration. Some progress, however, has been made. By 1984 twenty-four states had adopted statutes relating to artificial insemination using either the father's or a donor's sperm; most of these stated that the offspring from AI sperm donors are to be considered legally legitimate. Similarly, some states have adopted Section 5 of the federal Uniform Parentage Act, which stipulates that if a husband and wife jointly sign a consent form for artificial insemination, the husband is, for legal purposes, the same as the natural father.[20] Some states have also passed legislation regarding in vitro fertilization, though only Pennsylvania has legislation that goes beyond the question of experimentation and requires practitioners to keep records of the names of sponsors and users of this technology.[21]

Beyond these limited incidences, advances in policy for biomedical technologies as a whole, including genetic screening, have come from three sources: federal and state advisory boards, judicial decisions, and voluntary guidelines from professional organizations and from concerned citizen groups. The most significant federal advisory board has been the National Commission for the Protection of Human Subjects of Biomedical and Behavioral Research, established by the National Research Act of 1974 and in existence till 1978. Though primarily concerned with the rights of fetuses and persons used in research, the commission also published a study in 1978 on the "Implications of Advances in Biomedical and Behavioral Research." The study argued for the need to "create new institutions to monitor the development and introduction of new technologies in the biomedical fields, and to draw the attention of legislatures and the public to social problems arising from the use of those new technologies."[22] Furthermore, it pressed for establishment of a national agency to study and regulate new developments in biomedical research and technology and to formulate policy to inform the public of them. Such an agency has not yet been set up, although it is especially needed since the breakup of the Department of Health, Education, and Welfare and the dissolution of the Ethics Advisory Board, which had advisory responsibility for matters of biomedical research and implementation.

Judicial Policy

In the absence of such a specialized agency, virtually all existing policy has come from the courts. A myriad of court cases have been conducted at all judicial levels regarding each technology discussed in this chapter. Some of these cases have special and enduring import in the development and use of the new techniques. Certainly the most important case in the history of reproductive biomedical technology was the *Roe* v. *Wade* decision legalizing abortion in 1973. This case was crucial in the widespread deployment of amniocentesis, ultrasound and genetic scanning, and sex preselection. However, a second step was taken in 1977 in the case of *Planned Parenthood* v. *Danforth*. In this instance the court reaffirmed the position in *Roe:* that the "fundamental" decision to give birth belonged solely to the prospective mother in consultation with her physician. With this decision, artificial insemination and in vitro fertilization became the choice of a woman regardless of the wishes of her husband or any prospective father who had donated sperm.

Both the Roe and Danforth cases reached the U.S. Supreme Court, thereby making their decisions final in lieu of constitutional amendment. However, other technologies like surrogacy and genetic screening of either fetuses or adults have not benefited from Supreme Court rulings, even though lower courts have heard cases involving these technologies. Many of these cases are now on appeal. In addition, wrongful-birth suits are almost routinely filed against physicians who have failed to inform prospective mothers of the techniques available to detect birth defects that later appeared in children being carried. Many of these suits have succeeded, and they have encouraged the medical community to use screening and scanning techniques. Such medical encouragement constitutes a policy itself and one the courts have implicitly adopted in deciding cases in favor of the mother's suing her doctor for child support.

The number of cases in state and federal courts concerning matters of genetic and reproductive technology has burgeoned. These cases usually involve the rights of privacy upheld in *Roe* and *Danforth* and are decided on an ad hoc basis. Because biomedical technologies frequently raise questions of individual rights, the tendency toward court involvement in decision making is expected, particularly since the rights invoked are the fundamental ones of privacy, self-determination, or compelling state interest.

Though the role of the courts in biomedical policy is understandable, it is nevertheless disturbing for several reasons, some of which touch upon the very nature of the judicial process and upon its legitimate role in the development of national policy. The judicial system is by nature reactive; it cannot anticipate new policy needs or questions, and it is invariably slow in resolving disputes. Courts must wait until cases are brought to them before deciding general issues of right infringement. However, biomedical technologies are fast developing and need regulation or decision long before the time usually required to bring a case to court and to resolve it.

Also, the judicial method of resolving disputes is necessarily ad hoc—only one case is settled at one time. This is true even in light of the U.S. legal doctrine of stare decisis (the rule of precedent) because in biomedical cases many factors coalesce to restrain any judicial tendency to establish general precedents. The first factor is the judges' (or juries') lack of medical or scientific expertise. Judges are forced to rely on the testimony of experts, who are called to support both sides of the issue. The common disagreement among alleged experts results in a natural conservatism of judges: They tend to avoid extending the decision further than to the specific case being heard.

A second contributing factor is that the legal or constitutional principles involved in cases concerning reproductive technologies or genetic screening are themselves vague and subject to interpretation.

Thus the appellate success of any judge's decision is frequently doubtful even for the presiding judge himself. This ambiguity again encourages conservatism in the decision. Third, the judicial system lacks enforcement power of its own and must rely on other agencies to carry out its decisions. Because such agencies are often political, enforcement may depend on the popularity of the decision, and, as in the example of abortion decisions, unpopularity has resulted in lack of enforcement. Finally, even though most judges are not subject to electoral pressures, this impartiality does not mean that they are the best adjudicators of difficult policy problems involving inflammatory political issues like the rights to life, to privacy, or to a certain quality of life. The judicial community recognizes that it is in many instances not the ideal policy maker, and therefore judges often invoke the doctrine of judicial restraint to justify their refusal to render a precedent-setting decision. This doctrine is yet another conservative factor hindering authoritative policy making for biomedical technology by the courts.

Citizen Action Groups

It can be argued that no other arm of the U.S. policy-making system is faster or more willing than the courts to resolve the dilemmas of biomedical technology. The record (or lack of one) of policy making in Congress, the White House, and the bureaucracy would certainly bear this argument out, and for this reason nongovernmental organizations have sought to step in and fill the policy void. Associations ranging from those of the U.S. medical community to the various right-to-life organizations around the country have all entered the fray of making policy—with dubious policy outcomes.

In the absence of governmental action citizens would naturally seek to establish their own guidelines regarding the uses of biomedical technologies. In fact, it is in many ways reassuring to see the American Fertility Society and the American Association of Tissue Banks issuing their own guidelines for doctors and researchers involved with artificial insemination and fetal research. These instances recall the self-regulation of recombinant DNA researchers meeting at Asilomar (discussed in Chapter 7) and exhibit a high degree of concern with public safety. However, the legal status of regulations of even professional groups is negligible outside a court of law and is itself often a victim of political whims and fashions.

Furthermore, even though groups like Planned Parenthood and Right to Life express the opinions of many citizens and provide valuable service in raising the issues the public must consider, they do not have either the constituency or the mandate to make policies for an entire nation. The sheer number of such organizations allied on either side of the abortion and reproductive technology questions indicates the need for general, authoritative policy making by institutions of government. The issue of abortion and the attendant issues raised by the reproductive technologies have, in their proliferation of private-opinion organizations, gone beyond the realm of public discourse into dangerous areas of political schism and even violence. By early 1985, groups opposed to abortion and reproductive choice on either moral or religious grounds had destroyed over thirty abortion and reproductive-technology clinics. The need for policy has grown ever more insistent.

CONCLUSION

It can be argued that in the gamut of policy alternatives from prohibition to mandate, the absence of policy making is itself a conscious policy choice. This argument, which is often valid, has been vigorously revived by the Reagan administration in its advocacy of deregulation and government non-intrusion into people's lives. On its face, biomedical technology seems a policy area naturally amenable to this policy-by-omission approach because of the intrinsically personal and private techniques involved in reproduction.

However, reproductive technologies concern not only the living but the not yet born and even the not yet conceived. Although antiabortion activists frequently ask who speaks for the unborn's right to life, new reproductive technologies raise the question of who will speak for the unborn's right to a certain quality of life. The new technologies provide the knowledge at least to ask—and perhaps to answer—that question. The question of quality of life is intrinsically controversial to some because it raises the possibility of abortion perhaps even to the degree of genetic engineering to eradicate genetic disease. The spectre of wrongful-life cases hovers over the U.S. legal system precisely because no policy answers have been forthcoming to the query of whether we should encourage the use of reproductive technologies to ensure a certain quality of life to all newborns. The idea of bringing suit because one has been allowed to be born with a horrible genetic and fatal disease sounds ludicrous; yet in only one state (California) have such suits been legislated against or even legislatively considered.

Given the nature of reproductive technologies, perhaps government ought not to make policy that tells individuals how to use their bodies or exercise their right of choice in reproductive decisions. Perhaps every individual should be given those rights to exercise as she or he sees fit. Perhaps—but according to democratic theory, government in making policy decisions relies on individuals to exercise those very rights of choice. Whether the process is one of legislation, referendum, or constitutional amendment, democratic government by its nature relies on citizens to tell it what the decision will be. The right of choice in reproduction does not exclude policy making by relevant governmental agencies. Indeed, in the area of policy making for reproductive technology citizen choice is essential and guaranteed by the democratic process. The knowledge that reproductive technologies provide renders policy making difficult, but policy must be made. Government's refusal to make such policy not only threatens the reproductive rights of the living but also the same rights of the not yet born.

NOTES

1. Constance Holden, "Two Fertilized Eggs Stir Global Furor," *Science* 235, no. 4657, July 6, 1984, p. 35.

2. *Ibid.*

3. *Ibid.*

4. Robert H. Blank, *The Political Implications of Human Genetic Technology* (Boulder, Colo.: Westview Press, 1981), p. 39.

5. *Ibid.*, p. 63.

6. *Ibid.*

7. L. R. Kass, "Implications of Prenatal Diagnosis for the Human Right to Life," in J. M. Humber and R. F. Almeder, eds., *Biomedical Ethics and the Law* (New York: Plenum Press, 1976). Quoted in Blank, *op. cit.*, p. 30.

8. Blank, *op. cit.*, p. 47.

9. National Institutes of Health, *Antenatal Diagnosis: Predictors of Hereditary Disease or Congenital Defects* (Washington, D.C.: Department of Health, Education, and Welfare, 1979.)

10. However, the evidence also suggests that Americans care about the order of their offspring. Many adults polled preferred to have a male child first, followed by a female child.

11. Blank, *op. cit.*, p. 117.

12. See Blank, *ibid.*

13. National Academy of Science, *Genetic Screening: Programs, Principles, and Research* (Washington, D.C.: NAS, 1975), p. 9.

14. See Barbara Katz Rothman, *In Labor* (New York: W. W. Norton, 1982), chap. 4.

15. Barbara Katz Rothman, "How Science is Redefining Parenthood," *Ms* 11, no. 1 and 2 (August 1982), pp. 154–158.

16. Rothman, *In Labor, op. cit.*, chap. 8.

17. Blank, *op. cit.*, chap. 4.

18. Office of Technology Assessment (OTA), *Development of Medical Technology: Opportunities for Assessment* (Washington, D.C.: Government Printing Office), 1976; and OTA, *Assessing the Efficacy and Safety of Medical Technologies* (Washington, D.C.: Government Printing Office), 1978.

19. Blank, *op. cit.,* p. 158.

20. Andrea L. Bonnicksen, "In Vitro Fertilization, Artificial Insemination, and Individual Rights: A Review of Policy," paper delivered at the 1984 American Political Science Association Annual Meeting, September 1984.

21. *Ibid.,* p. 17.

22. Quoted in Blank, *op. cit.,* p. 159.

SELECTED READINGS

Beauchamp, T. L., and J. F. Childress. *Principles of Bio-Medical Ethics.* New York: Oxford University Press, 1979.

Bergsma, D., ed. *Ethical, Social, and Legal Dimensions of Screening for Genetic Disease.* New York: Stratton, 1974.

Birch, C., and P. Albrecht, eds. *Genetics and the Quality of Life.* Australia: Pergamon, 1975.

Blank, Robert. *The Political Implications of Human Genetic Technology.* Boulder, Colo.: Westview, 1981.

Ellison, D. L. *The Bio-Medical Fix.* Westport, Conn.: Greenwood Press, 1978.

Etzioni, A. *Genetic Fix: The Next Technological Revolution.* New York: Harper and Row, 1973.

Finegold, W. J. *Artificial Insemination.* 2nd ed. Springfield, Ill.: Charles C. Thomas, 1976.

Goodfield, J. *Playing God: Genetic Engineering and the Manipulation of Life.* New York: Random House, 1977.

Lipkin, M., and P. Rowley, eds. *Genetic Responsibility: On Choosing Our Children's Genes.* New York: Plenum Press, 1974.

Rowick, D. *In His Image: The Cloning of Man.* Philadelphia: Lippincott, 1978.

Veatch, R. M. *Death, Dying, and the Biological Revolution.* New Haven: Yale University Press, 1976.

9
POLICY MAKING FOR SCIENCE-BASED TECHNOLOGY

One premise in this book, illustrated by the preceding case studies, is that the twentieth century marks the beginning of a new era in the relations between science and technology. For whatever reasons, the abstract knowledge of science and the practical know-how of technology appear to be firmly linked. Along with this new union of science and technology come thorny problems for the policy maker: problems whose solutions call for a reevaluation of the desired relations between government, business, research, social values, and public participation.

In Chapters 2 through 8 we identified specific policy issues raised by new technological developments and reviewed the efforts of U.S. policy makers to cope with the consequences of science-based technology. In a field like nuclear power, policy makers have not yet successfully determined a policy for the disposal of radioactive waste. In other fields without established policy, such as those of reproductive technology and communications technology, the courts are forced to become policy-making bodies by settling suits on a case-by-case basis. The traditional job of the U.S. courts has been to interpret and apply policy, not to make policy in a piecemeal, ad hoc fashion.

Cases like those discussed in the previous chapters illustrate the weaknesses and inadequacies of the current science and technology policy-making process, as well as possible new mechanisms for improving that process. In the first section of this chapter we describe the special problems facing a modern policy maker. We find that their solution calls for a policy-making framework that can both accommodate a diversity of social and political values and yield a policy acceptable to the public. In the second section we examine the commonly used policy-making tool of risk-cost-benefit analysis to see whether it can deal adequately with diverse social values. In the third section we discuss proposals for combining the policy-making process with citizen input as a means for achieving publicly acceptable policy.

As we will see, the modern policy maker can resolve the problems that he or she faces only by taking a stand on some of the most fundamental issues of ethics and political theory. Our goal is not to resolve these fundamental issues but simply to demonstrate the importance of such issues to the process of making policy concerning science and technology.

POLICY PROBLEMS OF MODERN TECHNOLOGY

The union of science and technology has proved to be powerful and troublesome. Knowledge of the structure of matter has enabled us to release vast amounts of energy from the atom and to synthesize new types of chemical substances; but accompanying the utilization of this knowledge come risks of poisoning the environment today and for the distant future. Knowledge of the biological processes of reproduction and heredity enables us to produce new life forms and to exert greater control over the nature of our offspring; yet in applying this knowledge we risk introducing harmful life forms, and we raise moral questions about the rights of the unborn. Knowledge of electromagnetism enables us to transcend the previous limits of space and time for the storage and exchange of information, but new communications technologies present problems concerning the rights of privacy and equal opportunity.

Technological developments throughout history have had substantial—sometimes even devastating—effects on social life. The potential social impacts of the current synthesis of science and technology appear to be even greater and longer lasting than those of earlier technological developments. We believe that the technological revolutions of the twentieth century will have far-reaching consequences that affect the health, environment, and overall quality of life, as well as the economic opportunities for a large segment of society. These effects, many of which are irreversible, extend to all segments of society now and in the foreseeable future.

Because of the increased scope and heightened intensity of the effects of the new science-technology synthesis, the welfare of more people is at stake with any given technological development and a greater variety of deeply cherished values is involved. Furthermore, people are more likely today than in the past to be concerned with the potential impacts of technology and to take steps to block the implementation of a policy that they perceive would inadequately protect their interests.

Because modern technology is based on revolutionary, highly specialized knowledge, it poses special problems for the policy maker, who is forced to rely on the testimony of technical experts to ascertain the possible risks of employing a new technology. Yet the experts frequently disagree in their evaluations of the risks, in part because the relevant theories are new and revolutionary and hence many details still need to be worked out. The disagreement is also a natural result of applying abstract, idealized concepts to complex, real-world situations. No matter how objective experts try to be, they implicitly employ different values and may consider different factors to be relevant or may weigh these factors differently.

PROBLEMS OF CURRENT POLICY MAKING

We will now characterize more precisely the special problems facing a policy maker in the age of modern technology when evaluating the social consequences of implementing a technology and devising a policy that best serves the public interest. First, a policy maker must obtain the relevant technical information concerning possible benefits and risks. Although this approach sounds straightforward, knowing which expert to believe may be an insurmountable obstacle. Second, because modern technology interacts with a diversity of social values, a policy maker must somehow compare and combine values that are measured in such dissimilar units as dollars and cents, on the one hand, and long-term risks to health and the environment, on the other. Third, because the affected parties of a policy decision include many different segments of society or the entire public, a policy maker must consider the conflicting interests and values of a broad population. Critics such as R. E. Goodin and K. S. Shrader-Frechette have charged that the current policy-making process has failed to come to terms with these special problems of modern technology.[1] They claim that within the current policy-making framework of risk-cost-benefit analysis, which is described in the next section, policy proposals are evaluated primarily in terms of economic impacts and the value of free-enterprise capitalism, and policy controversies are treated as issues that will be resolved later when more technical information has been collected rather than as issues involving fundamental questions about values. In other words, the critics charge, policy makers ignore a whole spectrum of relevant social values and fail to acknowledge that technical experts may always reasonably disagree.

The problem with current applications of risk-cost-benefit analysis, according to Shrader-Frechette, is that most policy makers apply it from the perspective of a logical empiricist philosophy of science.[2] Logical empiricists, as we said in Chapter 1, believe that science yields objective knowledge that is value-neutral and that logic alone dictates a unique answer to any scientific question, given sufficient data. Logical empiricist assumptions, claimed Schrader-Frechette, enter into policy-making at two levels: Many policy makers assume that the information provided by technical experts is totally objective, and they assume that risk-cost-benefit analysis is itself a science. From the perspective of logical empiricism, disagreement among scientific experts indicates either that some experts have used faulty logic or that more data need to be collected, and the scientific status of risk-cost-benefit analysis implies that its results are objective and value-neutral.

As a result of logical empiricist assumptions, many policy makers see questions about values as irrelevant to science and technology assessment, and so they are insensitive to their own value assumptions. Furthermore, public controversy over policy proposals is seen as something that can be eliminated by simply informing the public of the facts. As we claimed in Chapter 1, philosophical and historical analyses support the position that there is no such thing as objective, value-neutral scientific knowledge. All scientific conclusions are based on some system of values that may vary with social context, including conclusions obtained through risk-cost-benefit analysis. Policy makers therefore need to be sensitive to the values that shape both their own conclusions and those of the experts, and they need to be aware of how these values interact with those held by the public.

In our opinion, a policy-making framework that is sensitive only to economic values would produce policy failures. If, on one hand, a policy maker ignores the special problems of modern technology, relying primarily on analysis of economic impacts, then policy may fail at the implementation stage, when citizens who believe that a policy unjustly ignores or even violates their rights may appeal to the courts and successfully block that policy. If, on the other hand, a policy maker considers the complex and difficult issues posed by modern technology, then policy may fail at the formulation stage because no other analytical framework is available that accommodates those issues. Concerned citizens may again appeal to the courts for a decision.

In our discussion of the cases in previous chapters, we have indicated that the current science and technology policy-making framework faces precisely these problems. The social issues raised by modern technology are real and concern many members of the public. Furthermore, the policy-making process has already failed in many areas, either because the public has blocked implementation or because no policy exists where one is needed. Recognition of the nature of the problem provides us with clues for its solution. First, we need a policy-making framework that acknowledges the diverse social values relevant to science and technology policy, and we need guidelines for combining these values in an overall assessment of policy alternatives. Second, we need a framework that results in policy acceptable to the public. Somehow the interests and concerns of the public must be ascertained and accommodated before a new technology is in place—in the earlier stage of policy formulation rather than in the later stage of policy implementation. We now consider whether the policy-making tool of risk-cost-benefit analysis is adequate in light of the social issues raised by modern technology.

RISK-COST-BENEFIT ANALYSIS

Advocates of risk-cost-benefit analysis claim that all rational, deliberate decision making has a certain logical structure, which is the same regardless of whether the decision maker is an individual making a personal choice, a corporate executive making a business choice, or a government official making a choice about public policy. In each case, the policy maker first considers the various courses of action, then evaluates the consequences of each alternative, and finally selects the course with the most desirable consequences. The specific information considered varies with the goals of the decision maker.

In our ordinary everyday decisions, typically made on the spur of the moment, this process may be cut short; however, decisions on public policy should be made only after a period of careful study and on the basis of duly scrutinized assumptions. To ensure adequate consideration, science and technology policy today is usually formulated by following a formalized, systematic decision-making process of risk-cost-benefit analysis. For example, environmental impact studies required by the EPA and technology assessments provided by NSF and OTA utilize this method of analysis.

In risk-cost-benefit analysis the decision process is explicitly broken down into several stages, and each stage calls for a specific type of analysis. Decision makers should then be better able to identify, acquire, and evaluate the relevant information. The six stages of analysis may be outlined as follows, although they cannot be pursued independently of each other.

1. Identify the problem and basic policy objectives.
2. Formulate alternative courses of action.
3. Identify relevant consequences of each alternative.
4. Assign a probability to each relevant consequence.
5. Assign a value, i.e., a numerical cost or benefit, to each consequence.
6. Combine the information obtained in stages 3, 4, and 5 and select the best alternative.

Critics of the current science and technology policy-making process do not dispute the value of using some kind of formal decision-making framework like that of risk-cost-benefit analysis. However, they assert that a problem arises concerning the assumptions policy makers typically make when applying this framework to science and technology issues. Because risk-cost-benefit analysis has been such a useful tool for economists and corporate executives, assumptions that were made when this framework was applied to private business issues are simply taken for granted by public policy makers and unwittingly carried over to the context of science and technology policy. Because these assumptions are unrecognized, critics claim, many scientists and policy makers act as if the application of risk-cost-benefit analysis were simply a matter of plugging the relevant facts into a given, unquestionable logical framework. They overlook the fact that each stage of risk-cost-benefit analysis, when applied to science and technology policy, embodies value judgments of social and political significance. When they simply take the approach of economists for granted, they adopt a narrow position on these value issues without even considering the alternatives. Thus it is critical that the traditional technique of risk-cost-benefit analysis be revised to yield an adequate policy-making framework for an entire society.

Values and Policy Options

Public policy makers act only when they perceive potential problems that require new policy or modifications in established policy. Once a problem is identified and understood, policy makers can begin to formulate types of policies that might lead to its solution. As they identify what developments warrant policies, what are the nature and scope of the problem, and what the objectives of policy should be, policy makers are influenced by their own value judgments: They are deciding what constitutes an area in need of policy and what objectives are important to public welfare. They also must determine when to act: Should they begin establishing policy when a product is in the research stage or only after it is on the market?

Such decisions embody prior judgments about the general objectives of science and technology policy. For example, suppose a policy maker is confronted with the so-called problem of nuclear energy. First he or she must define the problem; however, this definition is influenced by what the policy maker judges to be the primary objectives of policy formulation. If the policy objective is to protect the public's physical health only when a clear and present threat exists, then the problem of nuclear energy is defined as the regulation of pollution levels and the storage of radwaste for the nuclear power industry. In contrast, if the primary objective is to determine how a new science or technology is to fit within society, the problem of nuclear energy is defined as the determination of the course that would best coordinate nuclear energy with other energy types. In the second case, the problem area has been broadened to include not only the effects of nuclear pollution on human health but also the comparative effects of other energy sources. Furthermore, policy making would be required for the research on nuclear energy as well as for its actual use.

In addition to the initial description of the policy problem, other factors also constrain the choice among available policy options. Policy options must be politically realistic: They should be compatible with U.S. political ideology and political institutions; they must be enforceable; and they must not offend potentially powerful interest groups. In determining compatibility the policy maker must again employ value judgments: He or she must first decide what are the current tenets of U.S. political ideology and how idealized goals should be balanced with political realities. The line between politically realistic and politically unrealistic options is not always easily drawn. As political theorist Robert Goodin pointed out, using the word *impossible* may be just an excuse for preserving the status quo.[3] What initially seems impossible may in fact be attainable with more effort.

On the surface, using risk-cost-benefit analysis as a tool for reaching policy decisions appears to have no direct bearing on identifying policy problems and policy options. The connection that exists is the result of a historical relationship between technology and business rather than of principles intrinsic to this type of analysis. Since 1900, risk-cost-benefit analysis has been the primary analytical tool of economists and private business, and new technology has been developed largely by private businesses that fund basic scientific research only as a means to increased profits. These associations suggest that the techniques for applying risk-cost-benefit to science and technology have been developed by economists in light of the goals of free-enterprise capitalism. Since policy makers naturally define problems so that they can be solved by the available, proven methods, they tend to define policy problems for both science and technology in terms of minimum regulations for existing industries and to stress the objectives of meeting consumer demands and protecting profits.

Values and Relevant Consequences

Once a policy maker has identified a policy problem and formulated various policy options, he or she must identify relevant consequences of these options. Here the policy maker must determine which types of consequences might bear on the final policy decision; in other words the policy maker must choose a set of categories for describing the impacts of a policy. Value judgments would influence this choice: A policy maker must discriminate between impacts that might be socially harmful or beneficial and impacts that either would not concern society or would lie outside the domain of government.

As an example of the decisions mandated at this stage of analysis, let us consider siting policy for nuclear power plants. The location of a nuclear power plant may affect such categories as energy costs for industrial and residential consumers, insurance rates, local property values, the physical and mental health of local residents, the well-being of other species, and the beauty of the landscape. The policy maker uses a value judgment each time he or she labels the effects of the power plant on any category either good or bad. He or she must also decide if each category is relevant to public policy and thus germane to the decisions. Because impacts of these types affect groups of people differently, each initial category may be broken down into such subcategories as physical health risks to children, physical health risks to adults, economic impacts on the poor, and economic impacts on the middle class. A policy maker may further subdivide each of these subcategories on the basis of time by considering impacts over five years, over ten years, and perhaps even over a hundred years.

Operational Attributes. Because the number of possible categories for describing the policy impacts can be multiplied indefinitely, many experts on decision analysis recommend that policy makers choose only categories defined by operational attributes.[4] To be operational, an attribute or property must come with a scale for detecting increases and decreases, since in risk-cost-benefit analysis a benefit is defined as either an increase in a "good" property or a decrease in a "bad" property and a cost is defined as either a decrease in a "good" property or an increase in a "bad" property. Thus an operational category—because it is described by a measurable attribute—is one that supplies information that enables a policy maker to detect differences between alternative courses of action. In addition, some method must be available for ascertaining how the property is affected by a policy option.

To illustrate the nature of operational attributes, let us consider the category of economic impacts in nuclear power plant siting policy. Without further detail, this category is not operational because it is not described by an attribute that has a measurement scale. To make this category operational, we might subdivide it into the following subcategories: projected consumer demands for power versus energy-providing capabilities, projected costs of building the plant, and revenue brought to the local government. These subcategories are operational because (1) each comes with a scale for detecting increases or decreases, and (2) the effects of a policy option on each economic subcategory can be estimated on the basis of statistics. As another example, the category of environmental impacts is more difficult to make operational because less research has been conducted on environmental effects than on economic effects. However, we might consider the subcategories of radiation effects on human reproductive abnormalities, radiation effects on human cancer rates, radiation effects on the rate of wildlife reproduction, and radiation effects on the growth of vegetation. Each of these subcategories is operational because each comes with a scale for detecting increases, and presumably the effect of a policy option on each category can be estimated on the basis of laboratory studies.

Criticisms of Risk-Cost-Benefit Analysis. Critics of the use of risk-cost-benefit analysis in policy making charge that, because this method is intrinsically biased in favor of easily quantified economic considerations, many important social costs and benefits are simply ignored. Regardless of whether one agrees with this claim, some cases of energy policy and nuclear siting policy seem to exhibit this bias: The external costs of air pollution, noise pollution, and the inequitable distribution of risk have all typically been ignored in technology assessments.

The bias in favor of quantifiable economic impacts and the general criticism against risk-cost-benefit analysis rest on two conceptual mistakes. This bias occurs when the advice to consider operational attributes is misinterpreted as advice to consider only objective properties and also when operational attributes are mistakenly equated with properties typically measured by a numerical scale as opposed to a qualitative scale. With a numerical scale the possible levels of an attribute (i.e., where the attribute falls on the scale) are assigned specific numerical values, whereas with a qualitative scale the possible levels of an attribute are ordered only according to a greater or less than relation. An objective attribute is defined as a property that does not depend on the attitudes or perspective of any individual; the measured level of an objective property is independent of any person's feelings about it. Furthermore, objective attributes typically come with numerical scales so they provide a policy maker with especially useful information. Logical empiricists regard objective properties as the only properties appropriate in a scientific assessment of science and technology policy since only these properties correspond to hard, objective facts.

As an illustration, the projected monetary cost of building a nuclear power plant would be an objective attribute: The cost is either $3 billion or it is not, regardless of whether one approves of nuclear power plants. In contrast, the cost of a nuclear power plant to the beauty of the landscape qualifies as a subjective attribute: The level of the cost depends on an individual's attitudes toward the landscape. Similarly, the cost of a nuclear power plant to the mental health of a local resident is a subjective attribute since the level of this cost depends on that individual's awareness of the hazards and attitudes toward risks. Subjective attributes like these are not usually associated with numerical scales because the numbers assigned to beauty and mental health would be arbitrarily chosen. Nevertheless, these attributes can be measured in terms of a rough qualitative scale sensitive merely to increases and decreases in beauty or mental health. Information of this sort may be relevant and useful in comparing policy options.

The danger of confusing operational attributes with attributes that are objective or associated with a numerical scale is that relevant and potentially useful attributes may be ignored. Contrary to the claims of some critics, the practical need to describe consequences in terms of operational categories does not rule out the inclusion of either subjective or qualitative social costs and benefits. For an attribute to be operational, it need only be useful in comparing the costs and benefits of different policy options, and this condition can be satisfied by subjective attributes associated only with a qualitative scale. People's feelings about a situation may be relevant because they affect their quality of life; such information could be obtained through questionnaires. In fact, technology assessments that considered the qualitative values of beauty and clean air have been conducted, and these values were found to influence the final policy decision.[5]

Ultimately we must acknowledge that the current usefulness of a category is a function of how much study it has received in the past. Vague categories can be made precise through careful analysis, and scientists can create measurement scales. Because the study devoted to a category is also a function of its perceived importance, currently nonoperational subjective values may become operational if they are considered important enough to merit study. For these reasons, no category falls outside the scope of formal decision analysis simply as a matter of basic principle.

Values and Risk Assessment

Once a policy maker decides upon a set of categories for describing relevant impacts, he or she must attempt to predict the actual impacts of each policy option. Decision analysts distinguish between three types of decision-making cases. In the first type, the outcomes of each course of action are known with certainty and are assigned a probability of one. In the second type—called decision making with risk—the actual outcomes of some alternatives are not known with certainty, and instead possible outcomes are assigned some specific degree of probability between zero and one. In the third type—called decision making with uncertainty—the possible outcomes of each course of action are listed but are not assigned a probability; the likelihood of these outcomes is simply unknown.

The distinction between decision making with risk and decision making with uncertainty is important because the two types require different methods in making the final decision. When the probabilities have been assigned to all significant impacts, as they are in decision making with risk, standard procedure dictates that one use a rule known as maximizing expected utility. According to this rule, a decision maker selects a course of action likely to yield the most benefits versus costs over time. The decision maker is gambling, but the gamble is calculated since he or she knows the probabilities of all costs and of all benefits. In decision making with uncertainty, however, a decision maker does not know which alternative is likely to yield the most benefits versus costs over time. He or she knows what can happen but not the likelihood of various costs and benefits. In the face of such uncertainty, a decision maker may either play it safe and select an option that avoids extreme costs or he or she may select an option that could yield extremely high benefits.

In the context of science and technology policy making, the cases confronting the policy makers cannot be neatly classified as one or another type of decision case. Some immediate impacts of a policy option may be known with certainty: for example, monetary costs of installing pollution control devices for a given level and type of air pollution. Policy makers and the public alike are most concerned about the hazardous impacts on human health. Some of these impacts, like death caused by a given level of radiation, are assigned a specific probability, but the risks of many health hazards, because they are not known, fall within the domain of the uncertain. A policy maker must decide whether to approach such uncertain impacts as decision making with risk or as decision making with uncertainty. The choice requires a value judgment by the policy maker since it depends on the degree of importance attached to the uncertainties.

One strategy for eliminating uncertainty is to transform unknown risks into known risks through research. The most reliable method for assigning a probability to a future event is to use statistics for similar events in the past. Since this method requires that relevant data from past events be on record, it is not useful for new technologies or very rare events like a nuclear power plant meltdown; in these situations a sufficient number of similar cases has not occurred to warrant a prediction. Many science and technology policy decisions about events in new contexts or undocumented fields require the sophisticated but less reliable technique of theoretical modeling.

Theoretical Modeling. The aim of theoretical modeling is to assign a probability to a specific type of event given certain initial conditions. For example, the purpose of the 1974 Rasmussen report, commissioned by the AEC, was to assign a probability to the event of a nuclear power plant meltdown, given initial data about the construction and operation of these plants. Using accepted scientific laws and theories, experts construct possible chains of events that may develop out of the given initial conditions. These imagined scenarios are represented by a "fault tree": The initial conditions branch into several possible events, each of these events in turn branches into other possible events, and so on through a sequence of branching events. By assigning appropriate probabilities to each branch and by seeing how many of the scenarios end with the event of interest, probability analysts calculate the likelihood of that event given the initial conditions. In the case of the Rasmussen report, experts constructed possible scenarios that would result in core meltdown, and they assigned a probability of 1 in 17,000 per reactor per year to that event.

By 1980 the conclusion reached in the Rasmussen report was regarded as inaccurate.[6] Crucial physical factors like the deterioration and aging of a plant were ignored, and the emergency core-cooling system was assumed to be reliable. More important, nonphysical factors outside the domain of scientific theory—like human error and sabotage—were not considered. The failings of the Rasmussen report conclusion illustrate limitations of theoretical modeling in general. This method is reliable only to the extent that all relevant factors and their interactions are taken into account. However, theoretical modeling requires an expert to make simplifying assumptions so that the laws and theories developed under controlled laboratory conditions can be applied. Thus a chance always exists that factors relevant in the world outside of the laboratory have been ignored. Because of the intrinsic limitations of theoretical modeling, it may be better to treat calculated risks for human life as uncertain.

The probabilities of many impacts of science and technology policy cannot be assessed by using either statistical methods or theoretical modeling. Statistical methods require a broad data base, and theoretical modeling requires well-tested scientific laws and theories. These methods are rarely reliable for predicting long-term effects since too many unforeseeable factors may intervene, and they typically do not apply to intangible social, political, and psychological effects. Consequently, most impacts that raise distinctly ethical issues (issues relating to the social values that define the just and fair treatment of human beings), as opposed to technical issues, are uncertain.[7] To take these into account a policy maker can employ a second strategy for eliminating the uncertain—simply ignoring those impacts that either lie in the distant future or result from unpredictable human behavior. Statisticians justify this strategy on the grounds that in the long run an impact is just as likely to result in an unpredictable benefit as in an equal and opposite unpredictable cost; thus the possible effects tend to cancel each other. If over time costs appear to be exceeding benefits, then a new technology or a new policy can remedy the situation.[8]

Principle of a Probability Threshold. To simplify the assessment of risk, policy makers sometimes use a third strategy known as the principle of a probability threshold.[9] According to this principle, any impact whose cost is assumed to be less than that of one death per year out of a million people may be ignored. The Rasmussen report concerning the risks of nuclear power plants, for example, stated that "an accident fatality risk to the public of . . . 10^{-6} or lower is considered negligible."[10] A possible rationale for using a probability threshold could be stated as follows: In the framework of risk-cost-benefit analysis, if the cost of a given impact is known as well as its probability, then the net worth of that impact is equal to its cost multiplied by its probability. Consequently, the rationale continues, a possible cost with the low probability of one out of a million per year does not have much net worth. Furthermore, the amount of money needed to reduce risk below the level of one death per million people per year probably outweighs any possible benefits. Anyway, according to another rationale, since the public already voluntarily accepts risks at this level, these risks might as well be ignored. (See Chapter 5 for a discussion of the fallacies behind this last argument.)

The strategies of ignoring impacts that either are unpredictable or fall below a certain level of probability may be legitimate in some cases. According to the rationale behind these two strategies, they apply only when a cost or benefit can be balanced by an equal, opposite effect. These strategies may be appropriate when the only concern is money; however, their applicability is questionable when such factors as death, disease, or fairness to future generations enter into the equation. Perhaps according to standards of ethics these types of harm cannot be balanced by benefits of any sort, so they cannot be ignored even though they are uncertain or unlikely. Sensitivity to this ethical issue does not entail abandoning risk-cost-benefit analysis; instead it requires improved criteria for identifying those impacts that can be ignored. It also requires acknowledging that policy decisions are intrinsically decisions with uncertainty.

Comparing Costs and Benefits

Once the actual impacts of all policy options are predicted as precisely as possible by whatever method chosen, a decision maker ranks the options in the order of their overall desirability, typically by computing a total benefit-to-cost ratio for each alternative. Regardless of the precise type of calculation employed by a policy maker, his or her final goal is to choose an option with the highest net benefit.

Before policy makers can rank the various policy options, they must address a difficult theoretical issue. Ranking options in terms of total benefits versus total costs requires comparing different types of benefits and weighing them against various kinds of costs. Regardless of the categories selected for describing impacts, it is highly unlikely that any one policy option would maximize all possible benefits while simultaneously minimizing all possible costs; instead, benefits in one category would accompany costs in another category. For example, decreased pollution levels usually accompany increased consumer costs, and present-day benefits of nuclear energy come with future costs of radwaste—what is good for one segment of the population may be harmful for another segment. Policy makers are thus required to trade off one type of benefit for another type of cost and to trade off benefits expected for one group of individuals against costs to other groups.

The problem posed by trade-offs can be described more precisely as the determination of a common scale for all categories of impacts. The necessary scale would be used to evaluate the worth of an impact, not to measure the objective level of that impact. Suppose, for example, that we already know that a given nuclear siting policy would yield, on the average, one death for every million local residents per plant per year and would cost a projected $0.09 per kilowatt hour of electricity for residential consumers. In contrast, a second nuclear siting policy would yield, on the average, two deaths for every million local residents per plant per year but would have a projected cost of $0.07 per kilowatt hour of electricity. These two possible categories of impacts—deaths per million and costs per kilowatt hour—qualify as objective categories because their actual levels are a matter of physical fact. However, to compare the overall desirabilities of the two policy options we need a common scale of worth for death rates and electricity rates; without this scale, it is difficult to balance the two types of impacts against each other. Even though we began with purely objective categories of impacts, in the final analysis we would enter the murky realm of value judgments.

Method of Compensating Variations. The usual technique for developing a common scale of worth for different categories of impacts is known as the "method of compensating variations," by which all gains and losses are converted into monetary amounts.[11] Using this method to determine the worth of a benefit we imagine an individual confronted with a sequence of choices: He may pay $1,000 for some specified degree of cleaner air or keep his money; he may pay $500 for this benefit or keep his money; he may pay $100 for this benefit or keep his money, and so on. In one option a person would be indifferent between keeping his money and having the cleaner air; this option is said to define the monetary worth of that benefit for that individual. Similarly, the worth of some specified risk for a person is defined as the minimum amount of money that person would accept as compensation for the given risk.

In principle, the method of compensating variations could be used to consider the preferences of each member of society in terms of his or her personal standards of value. The total worth of a policy option could then be computed by summing over its worth for each individual. In practice, however, decision makers usually assign only one fair market value to a given cost or benefit. The fair market value—the amount of money the public is willing to pay—is estimated on the basis of past social practice or social science surveys. (These methods of revealed preferences and expressed preferences are discussed in Chapter 5.)

Discounting. In addition to the practice of using fair market values, policy makers typically use a method known as discounting.[12] Discounting is a business technique for converting future monetary gains and losses into current monetary values. Because of inflation and the compounding of interest for investments, a gain of $100 today can be regarded as worth more than a gain of $100 ten years from now. Similarly a loss of $100 today would be valued higher than a loss of $100 ten years from now. To convert future monetary gains and losses into current monetary values, economists simply multiply projected amounts by an appropriate percentage. Some policy makers use the same discounting procedures for weighing future social costs and benefits against those of the present. As a result, future impacts count for less than more immediate impacts.

Pros and Cons of Methods. Although the standard methods of compensating variations and discounting may be appropriate in a private business context, critics of the techniques argue that they are inadequate in the context of public policy because they are insensitive to the ethical dimensions of policy decisions. Robert Goodin provided some striking examples of the implications of discounting at a rate of 5 percent per year.

> We would have to be indifferent between a clinic that cured 10 patients of a disease immediately and a preventative health system costing as much that prevents 16 people from contracting the same disease in ten years time. . . . Or, again, we should be indifferent between killing one person today with the sulphurous emissions from coal-fired power plants and killing 1,730 with the leaking wastes from nuclear power plants in two hundred years time.[13]

Applying the method of discounting when determining a policy, critics argue, presupposes that we have no special responsibility to future generations. The fact that they would have to deal with problems that they neither chose nor created carries no extra weight. According to critics, discounting also presupposes that the welfare of future men and women has less value than our own current welfare, although we have no reason to suppose that people of the twenty-first century would feel any differently about death and disease than do we of the twentieth century.

Using a free market value as the measure of a social cost or benefit is also ethically problematic. The impact of a social cost upon a person frequently depends on his or her current level of well-being. Environmental pollution, for example, is especially dangerous to those already in poor health. Exposure to radiation may cause life-long problems for children, which adults would not face. The financially poor are especially likely to be adversely affected by environmental hazards: They cannot afford to move away from a polluted area, and they are less able to pay for any resulting medical costs. Because free market values simply are not sensitive to people's differing needs, policy using them is likely to be biased against specific classes of people.

The ethical issues raised in connection with free market values and discounting can be accommodated within the framework of risk-cost-benefit analysis. The framework does not require that future impacts be discounted; in fact, a mechanism that uses reverse discounting can be employed to give more weight to future impacts. Furthermore, the objections to assigning a free market value to social costs and benefits could be answered by distinguishing more carefully between different groups of people and by assigning monetary values on the basis of group membership.

Other questionable aspects of the method of compensating variations are less easily dismissed. First, the method rests on the assumption that the concept of social welfare is adequately defined in terms of a monetary scale. However, everyone does not attach equal importance to $10.00; the evaluation depends on a person's income level, financial need, and life-style. Even though two people may both say that they are willing to pay $10.00 for a reduced risk of some disease, this agreement does not necessarily mean that they attach the same value to the reduced risk. An even more fundamental assumption is that loss of a human life or long-term suffering caused by disease can be balanced by some amount of money: According to some ethical systems, these costs may be beyond any price. In such ethical systems the preservation of life and the elimination of avoidable suffering are regarded as absolutes that are not subject to negotiation.

A second questionable aspect of the method of compensating variations is the implicit assumption that individuals regard the real worth of a gain or loss as simply what they would gain or lose themselves. Because it rests in part on this assumption the method of compensating variations may be said to beg some basic ethical issues: It assumes that people are basically selfish, that any one personal system of values is as valid as any other, and that the public good is defined in terms of what the majority prefers. The opposing assumptions could be defended equally well.

Decision theorists and policy makers are not oblivious to the limitations of the method of compensating variations. In the context of science and technology policy, some method for comparing different types of costs and benefits is necessary. Trade-offs are a fact of life.

But we also need new techniques that overcome the limitations of past methods and that embody the ethical principles we wish them to embody. Some decision theorists recommend developing a non-monetary numerical scale that measures the worth of an impact to an individual.[14] Others recommend a system of partial quantification by which some impacts are evaluated in terms of numerical market values whereas others receive only a qualitative assessment.[15] Each alternative can be made to fit a range of ethical values through the use of weighting factors. Preferences of the poor, for example, could be given twice as much weight as those of the rich, and preferences concerning health could be given one thousand times the weight of preferences concerning consumer electricity rates. To actually select the numbers for these weights the decision maker would have to consider carefully the ethical implications of preferences.

Ethical Theory and the Best Alternative

The entire formal decision-making process is geared toward the final point of selecting the best policy option. Within the standard framework of risk-cost-benefit analysis, the rule guiding this final selection is to pick an option with the highest numerical value for total expected benefits minus total risks summed over all affected individuals. To use this rule in formulating public policy the decision maker presupposes a specific type of ethical framework known as utilitarianism, which is associated with the well-known slogans of "maximize the expected utility" and "the greatest good for the greatest number." In this context "the greatest number of people" does not necessarily mean majority rule; it simply means that the welfare of all concerned parties should be considered equally.

Criticisms of Risk-Cost-Benefit Analysis. Critics of utilitarianism challenge the appropriateness of using standard risk-cost-benefit analysis as a tool for determining public policy. The critics feel that utilitarianism, and thus risk-cost-benefit analysis, ignores considerations of justice. According to the utilitarian selection rule, the critics argue, an option is ranked as best if it involves the greatest amount of total benefits versus total risks without regard to how these benefits and costs are distributed over a society. If a policy maker were to adopt the so-called best policy, all the benefits could accrue to one small class of society and all the costs could fall to the very large remainder. As long as the benefits to a few were great enough to outweigh the small costs to the many, the critics point out, such a policy selection would be ethically permissible. In a more likely case, every member of a society would receive some benefits and have to pay some costs. However, because utilitarianism requires that equal consideration be given to each individual, such a policy could discriminate systematically against minority groups with special interests or needs. The critics are not charging that risk-cost-benefit analysis is always unfair: rather that it does not adequately protect against possible injustices. (In Chapter 4 on energy policy, we found that this charge is often justified.)

The ethical absolutists raise a second objection to any type of risk-cost-benefit analysis in the context of public policy. Ethical absolutists are defined as people who maintain that the morality of an action does not depend upon the consequences it happens to cause and that some types of actions are morally unjustified no matter what benefits might ensue from them. They thus disagree with the utilitarian idea that any cost or any benefit is subject to trade-offs: According to the absolutists, some costs—typically those that violate basic human rights like the right to life and to liberty—can never be balanced by other considerations. For example, they believe that policy makers cannot morally justify the loss of one life on the grounds that they have saved two, and they cannot justify any involuntary risk of death, no matter how small. Absolutists charge that utilitarians confuse practicality with morality—utilitarians reply that absolutists are irrational to sacrifice the possibility of much good for the sake of avoiding a lesser evil.

Obviously the debate between utilitarians and absolutists cannot be resolved here; in the context of public policy, each side has some merit. According to a middle position, policy makers should be sure that their approach incorporates ethical concerns and does not equate ethics with mere practicality. However, in some situations a policy maker may have no alternative but to make trade-offs; sometimes every option available, including that of taking no action, violates some cherished right. Limitations on liberty sometimes accompany the avoidance of deaths, and the freedom of some people may be purchased only at the expense of the freedom of others. Thus, on one hand, a policy maker would be unwise to abandon completely the framework of risk-cost-benefit analysis; on the other hand, he or she may find that it is necessary to consider seriously some alternatives to the rule of maximizing expected utility.[16]

Alternative Approaches. The ethical philosopher John Rawls has proposed a decision-making scheme that avoids some of the objections directed against utilitarianism.[17] According to Rawls, a policy maker would proceed through all six stages of risk-cost-benefit analysis as described in section 2, taking special notice of the effect of policy options on different groups of people. By taking into account this information, the policy maker may then consider that the most just option is the best policy choice. This option would be the one that accorded the greatest benefit to the class currently worst off, as long as people in that class remained the worst off even after receiving this benefit. If no such class was involved, or no such option was available, the policy maker would use the rule of maximizing expected utility. According to Rawls's selection rule, a basic policy objective is equal distribution of benefits across society.

A second alternative to the rule of maximizing expected utility is known as the maxi-min rule, and it is one of the recommended rules for decision making with uncertainty. As we saw earlier in this chapter, some critics of risk analysis claim that all policy decisions should be treated as decisions with uncertainty: The degree of risk is too uncertain and the possible costs are too great to warrant a calculated gamble. According to the maxi-min rule, the best policy option is the one associated with the least possible harm. A policy maker may always choose the policy with the least risk to health or life, regardless of whether other options are associated with greater possible benefits. Supporters of this approach argue that the purpose of public policy is to protect society from certain "evil" impacts of science and technology—in this way some of the objections of the absolutists are met.

Other selection procedures are also possible for the decision maker. One technique is to choose only policies with reversible impacts[18] so that if a mistake is later discovered, it can be corrected. In this way a policy maker would not permanently burden future generations with current mistakes. Another approach is to use a priority list of minimum goals. For example, a policy maker may first consider only those options that meet a specified goal for environmental pollution, and then out of this class he or she may consider only those options that do not infringe upon the value of free choice. By following this approach the policy maker sets aside some types of impacts as being not open to negotiation in terms of other types of impacts. All the alternative procedures and rules considered in this section can be combined in various ways and with the rule of maximizing expected utility. Each possibility, however, carries ethical presuppositions that merit careful examination.

As the discussion in this section has shown, making an informed policy decision about science and technology is a complicated and difficult process. The quantity of necessary technical information is so overwhelming that the policy maker must be tempted to simply plug the claims of scientific experts into a ready-made computing model of analysis. But, as we have also seen, the computing machine for standard risk-cost-benefit analysis may require new, more up-to-date software. The traditional framework of risk-cost-benefit analysis embodies assumptions that in the context of modern science and technology policy become questionable ethical presuppositions. As a result, many of the perceived failings of current policy to deal adequately with public concerns may result in part because the current tool of risk-cost-benefit analysis embodies a system of values that is not in harmony with that of the public. Part of the solution rests with a careful reexamination of the value judgments implicit in each stage of decision making.

POLICY MAKING AND PUBLIC PARTICIPATION

In a democratic society, the interests of each member deserve consideration on matters of public importance: This is a "given." The job of democratic institutions is to ensure that the interests of all people are justly considered. For these reasons, U.S. citizens have a right to be heard on matters of science and technology policy, and the professional interests of government policy makers should be to listen carefully. Policy makers who overlook the concerns of their public may find themselves out of a job; at the very least they run the risk that their policies would be ineffective because of public opposition.

We have seen through the review of U.S. science and technology policy in this book that from both an abstract and a practical democratic perspective, the implementation of the current policy approach has not been totally successful. Our discussion of the decision-making tool of risk-cost-benefit analysis indicates a possible reason for this: Each stage of policy decision making—from the initial identification of basic policy objectives and problems to the final choice of a policy—requires the policy maker to make political, social, and ethical value judgments. For policy making to be successful, these judgments must somehow be made consonant with the values and interests of the U.S. public. Within the framework of democratic institutions, the best available way to achieve this goal appears to be to increase representative citizen input. But to do this the policy maker must address some difficult questions: How much influence should the public have in science and technology policy decisions? What type of political process best resolves the heated controversies that arise out of clashing interests and values?

Extreme Democratization of Policy Making

The most extreme approach to public participation in science and technology policy would be for the public to undertake all stages of decision making. Any group of citizens could identify a policy problem, policy options would be solicited from the public, and options would be subject to a binding federal, state, or local referendum. Each voter would be supplied with relevant information about possible benefits and risks to all segments of society and would be free to believe the experts of his or her choice and use his or her value system in selecting a policy.

Subjecting policy options to a public vote might be defended on two grounds. First, because science and technology affect the welfare of all members of society, each deserves some direct voice in setting policy. Second, because each stage of policy making involves ethical judgments about the welfare of society and the just treatment of individuals and because there are no technical experts in the field of ethics,[19] all mentally competent voters are equally qualified to make the required decisions.

Although direct participation may be defended as legitimate, critics argue that it is not necessarily wise.[20] Voters would be unable to digest all relevant technical information and translate it into personally meaningful effects nor could they be made aware of all the complex ethical issues posed by policy decisions. Some political theorists claim that the U.S. public is generally a poor judge of its own welfare.[21] In the case of determining science and technology policy, critics argue, the voters are not up to the task of making such complex decisions nor do they want that responsibility—the job is better left to expert policy makers.

Conservative Approach to Policy Making

In a more conservative democratic approach to citizen involvement in science and technology policy making, citizens would act as sources of relevant information to be used by governmental policy makers at their discretion. The public is currently treated as such an information source. Social scientists inform policy makers about some public preferences and attitudes toward risks; legislative hearings provide organized public interest groups with opportunities to air their concerns; citizens representing the public serve on advisory boards of many governmental agencies. In all of these cases attitudes of the public are relayed to governmental policy makers, but ultimately all decisions about the public good rest in the hands of these officials.

The conservative approach to public participation in science and technology policy has typically suffered from two types of problems: obtaining adequate information (1) about many different groups of people and (2) about a broad spectrum of social concerns. Social science surveys deal only with already targeted impacts on the "average" person. Legislative hearings tend to be dominated by well-organized groups with money and to focus on only one aspect of a policy issue such as environmental pollution. The few lay citizens on an advisory board are usually overshadowed by the overwhelming majority composed of technical experts concerned with factual questions.[22]

Moderate Approach to Policy Making

A practical solution to the political problems faced by modern science and technology policy rests somewhere between the paternalistic policy-making approach of the past and the opposite extreme of direct and complete public control. Our discussion of risk-cost-benefit analysis suggests ways in which a compromise might be achieved. As a first step policy makers might reevaluate the ethical presuppositions behind policy selection rules and rethink basic policy objectives. At this level public input is especially relevant: Citizen advisory boards that include a wide range of informed and thoughtful lay people could be formed. The boards might benefit from expert input in the fields of ethical theory and the history and philosophy of science and technology. Through these advisory boards the public could influence the basic framework of policy decisions. As a second step, the policy makers might devise a procedure for closer government and citizen cooperation in the actual formulation and evaluation of policy proposals. Rather than limiting its input to a particular, narrowly defined issue, the public could help define policy options, relevant categories of impacts, and worth of these impacts.

Achieving a harmony between science and technology and the interests and concerns of the public is one of the great challenges of the twentieth century. To meet this challenge governmental policy makers must acknowledge the intrinsic ethical and social dimensions of science and technology policy issues and thus expand the range of relevant "experts" beyond scientists and engineers to include members of the general public. At the same time the public must acknowledge the social and ethical significance of scientific and technological developments and prepare for its new role as an "expert" by being informed about the relevant types of consequences. Democracy flourishes only through the joint efforts of a society and its government.

NOTES

1. Two critics whose claims we will examine are Robert E. Goodin, *Political Theory and Public Policy* (Chicago: University of Chicago Press, 1982), and K. S. Shrader-Frechette, *Science Policy, Ethics, and Economic Methodology* (Dordrecht, Holland: D. Reidel, 1985).

2. Shrader-Frechette, *op. cit.*, pp. 67–74.

3. Goodin, *op. cit.*, chap. 7.

4. See Ralph L. Keeny and Howard Raiffa, *Decisions with Multiple Objectives: Preferences and Value Tradeoffs* (New York: Wiley and Sons, 1976), chap. 2.

5. For a discussion of specific cases, see Shrader-Frechette, *op. cit.*, p. 141.

6. See K. S. Shrader-Frechette, *Nuclear Power and Public Policy: The Social and Ethical Problems of Fission Technology* (Dordrecht, Holland: D. Reidel, 1980), pp. 82–89.

7. Goodin, *op. cit.*, pp. 188–193.

8. See Goodin, *op. cit.*, chap. 9, for a more detailed discussion of this principle.

9. See K. S. Shrader-Frechette, *Risk Analysis and Scientific Method* (Dordrecht, Holland: D. Reidel, 1985), chap. 5.

10. *Ibid.*, p. 128.

11. See K. S. Shrader-Frechette, *Science Policy, Ethics, and Economic Methodology*, chap. 5.

12. Goodin, *op. cit.*, pp. 179–183.

13. *Ibid.*, p. 182.

14. See Keeny and Raiffa, *op. cit.*

15. The system of partial quantification is discussed by K. S. Shrader-Frechette in *Science Policy, Ethics, and Economic Methodology*, chap. 6.

16. This point is made both by Goodin, *op. cit.*, and by Shrader-Frechette in *Science, Policy, Ethics, and Economic Methodology*.

17. See John Rawls, *A Theory of Justice* (Cambridge: Harvard University Press, 1971), for an entire scheme of principles governing just choices.

18. Goodin, *op. cit.*, p. 209.

19. Shrader-Frechette, *op. cit.*, pp. 302–304.

20. *Ibid.*, pp. 304–308.

21. This argument is considered by Goodin, *op. cit.*, chap. 3.

22. K. Guild Nichols, *Technology on Trial: Public Participation in Decision-Making Related to Science and Technology* (Paris: Organization for Economic Cooperation and Development, 1979), p. 58.

SELECTED READINGS

Giere, Ronald N. *Understanding Scientific Reasoning*, 2nd ed. New York: Holt, Rinehart and Winston, 1985.

Goodin, Robert E., *Political Theory and Public Policy.* Chicago: University of Chicago Press, 1982.

Nelkin, Dorothy, ed. *Controversy: Politics of Technical Decisions.* Beverly Hills, Calif.: Sage Publications, 1979.

Nichols, K. Guild. *Technology on Trial: Public Participation in Decision-Making Related to Science and Technology.* Paris: Organization for Economic Cooperation and Development, 1979.

Shrader-Frechette, K. S. *Science Policy, Ethics, and Economic Methodology.* Dordrecht, Holland: D. Reidel, 1985.

Singer, P. *Practical Ethics.* Cambridge: Cambridge University Press, 1979.

Smart, J.J.C., and B. Williams, *Utilitarianism, For and Against.* Cambridge: Cambridge University Press, 1973.

Stokey, Edith, and Richard Zeckhauser. *A Primer for Policy Analysis.* New York: Norton, 1978.

POSTSCRIPT

On January 28, 1986, millions of Americans watched with shock and horror as the space shuttle *Challenger* exploded shortly after takeoff, killing all seven crewmembers. Never before had such a tragedy occurred in the U.S. space program. Indeed, despite the obvious risks of space travel, NASA's enviable safety record made this event almost unthinkable to most people outside the space technology community. Years of glorious successes and technological triumphs in space had inured us to the unique—but seemingly surmountable—risks of reaching out to explore our celestial surroundings. On this day our reach was abruptly and terrifyingly cut off after less than two minutes and only ten miles above the earth.

The *Challenger* disaster carried with it some special ironies that heightened its sorrowful impact. The crew on this shuttle mission had seemed a demographic microcosm of U.S. people: two women—one of them Jewish—a black man, a man with an Oriental surname, even a man named Smith. Every person in the United States had a special symbol for whom to mourn. Even more disturbing was the presence of the first civilian and scientific layperson aboard the shuttle. Sharon Christa McAuliffe was a school teacher from New Hampshire, chosen by NASA from a pool of ten thousand applicants who responded to President Reagan's 1985 announcement of the first teacher-in-space program. In winning her place among the shuttle's crew, McAuliffe had stressed that her presence would allow millions of school children and ordinary Americans to participate directly in the technological conquest of space. For these vicarious participants, her death became a personal agony.

The tragedy of *Challenger* was actually a story of two modern technologies and of the dangers of participating in their use. Television coverage of the launch allowed millions to witness the disaster even before the NASA mission controllers in Houston saw it on their monitoring instruments. Through the immediacy of live-television technology viewers were forced to acknowledge the risks of space technology—risks forgotten or hidden beneath the nationalistic pride of seeing astronauts walking on the moon and returning safely to earth. And once again television demonstrated the psychological risks posed by its capacity to bring the world into U.S. living rooms. As in the cases of political assassination, Vietnam body counts, and hostage crises, the medium forced viewers to live and relive—through countless reruns of the videotape—the experience of violent death.

In the aftermath of *Challenger,* political leaders and policy makers will have to make many difficult decisions concerning the future of the U.S. space program. They will have to weigh the numerous intellectual, business, and military benefits of an active space program against its now very real and acknowledged risks and human costs. As this book goes to press in 1986, some in Congress have already called for an end at least to the practice of allowing civilians aboard future shuttle missions. No decisions have been made as yet either by Congress or NASA, other than to cancel all future shuttle flights until spring 1987. But one of the many hard lessons of this sad episode is that, in one way or another, modern technologies have made all citizens participants in their use and in their consequences. Therefore, we must all take part in the decisions that shape the future of these technologies, for it is our future as well.

INDEX